Institution of Engineering and Technology

IET and FSA International Semiconductor Forum 2007

May 14-15, 2007
Paris, France

Printed from e-media with permission by:

Curran Associates, Inc.
57 Morehouse Lane
Red Hook, NY 12571
www.proceedings.com

ISBN: 978-1-60560-083-3

Some format issues inherent in the e-media version may also appear in this print version.

Institution of Engineering and Technology

IET and FSA International Semiconductor Forum
2007

TABLE OF CONTENTS

Consolidation and the IP Business Model 1
No Author Given

Effective Chip Packaging….. Starts with Design 32
A. Longford

RF System in Package - Technology choice and Design Methodology 39
C. Barratt

Technology Aware Design 44
B. Dierickx

Smartphones and Beyond 65
D. Wood

The Benefits of Drawbacks of 90nm and below in Wireless 101
R. Nave

Consumer Electronics and the Semiconductor Industry 109
J. Yu

Asynchronous Circuits Adopted 148
A. Peeters

Winning the Supply Chain Challenges in a Fabless Model 175
R. Torten

DFY, DFT and DFM: A Pragmatic Approach to Built-in Quality 192
G. Curren

Panel: DFT, DFY and DFM: Ways to Build Quality from the Outset 198
No Author Given

Design for Yield, Design for Manufacturing and Design for Test 205
G. Fischer

DFY, DFT and DFM: Ways to Build Quality from the Outset 217
L. Tissot

Consolidation and the IP Business Model

ARM®

THE ARCHITECTURE FOR THE DIGITAL WORLD®

The Landscape...

- Semiconductor industry:

  - Fewer FABs
  - Global economic competition
  - Fabless model evolving
  - Fewer IDMs
  - Big names concentration

  - Is the IC industry maturing?
  - Is silicon technology slowing down?

  - Things are changing.......

New Business Opportunities

- Technical progress brings a basis for industry evolution
  - Miniaturization, reductions in chip costs, increases in complexity
- Increases complexity has exponential effect on design costs
  - Rising costs give way to specialisation and outsourcing

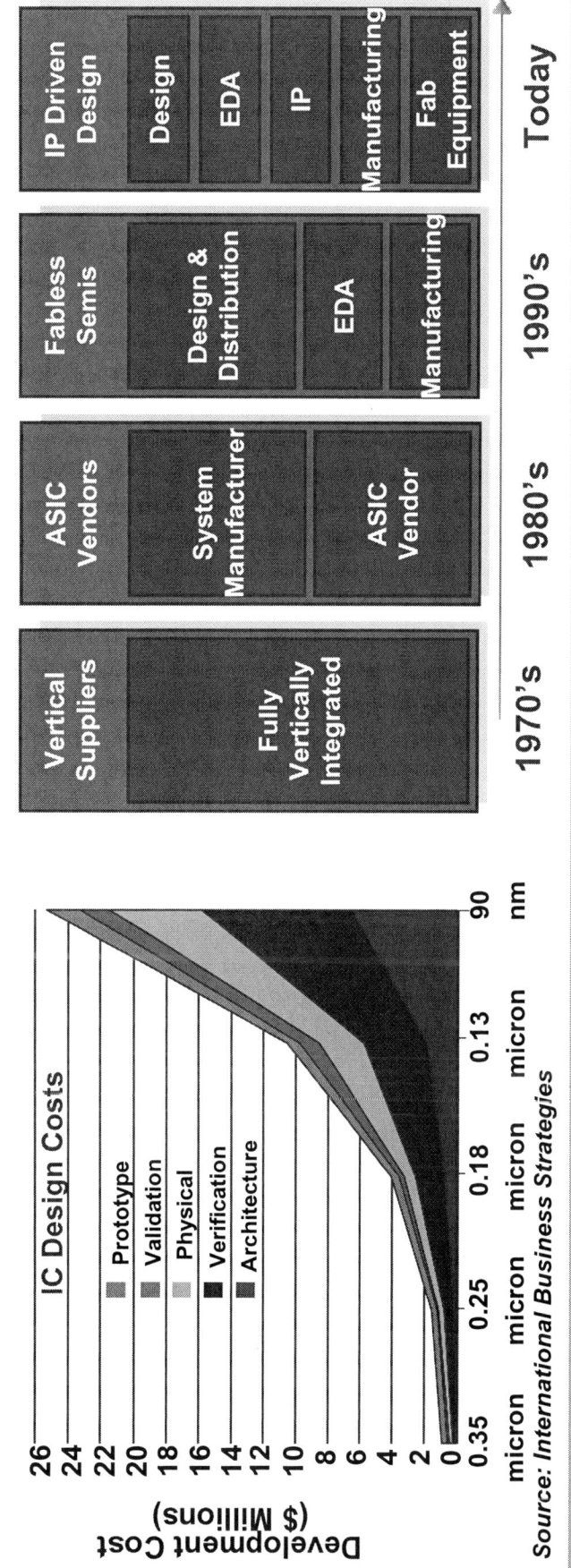

Gearing and Private Equity

Share price performance of UK companies with a higher debt/EV relative to UK companies with a lower debt/EV

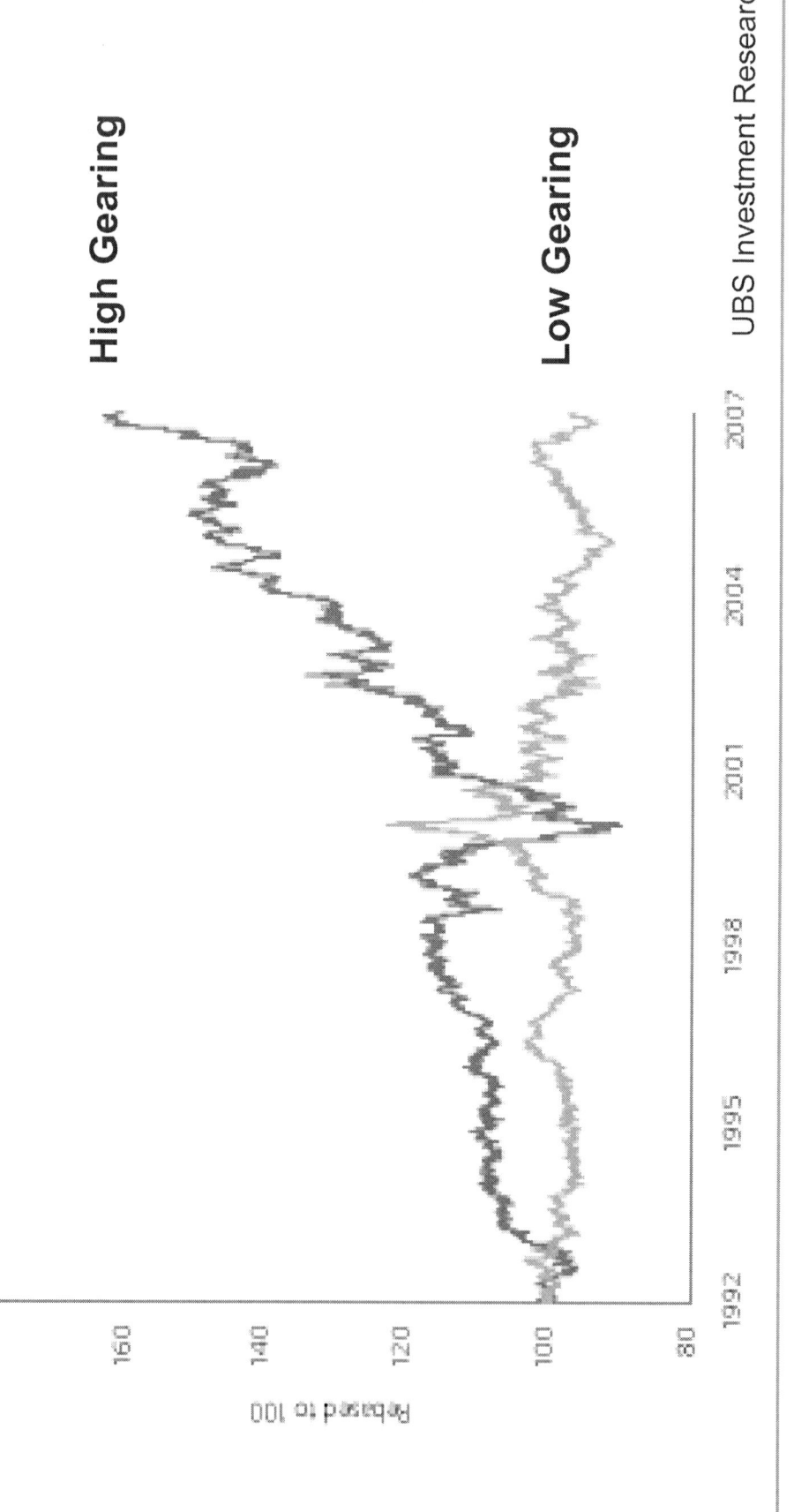

UBS Investment Research

Private Equity takes off in Semi

- Half a Billion $
  - 1997: Citicorp Venture and Credit Suisse buy Fairchild for $550 million
  - 1997: Texas Pacific Group buys Zilog for $527 million
  - 1999: Citicorp Venture and Credit Suisse buy Harris Semifor $520 million

- One Billion $
  - 2004: CitiGroup, Francisco & CVC pay $828 million for part of Hynix
  - 2005: KKR & Silver Lake buy Agilent Technologies' semi for $2.7 billion
  - 2006: Bain Capital buys TI sensors-and-controls for $3 billion

- Ten Billon $
  - 2006: KKR and Silver Lake Partners buys Philips Semi for $10 billion
  - 2006: Blackstone Group and TPG buys Freescale Semi for $17.6 billion

Who wins in 2015

1955	1965	1975	1985	1995	2005
Hughes	TI	TI	NEC	Intel	Intel
Transitron	Motorola	Fairchild	TI	NEC	Samsung
Philco	Fairchild	Philips	Motorola	Toshiba	TI
Sylvania	GE	NSC	Hitachi	Toshiba	Toshiba
TI	RCA	Intel	Toshiba	Motorola	ST
GE	Sprague	Motorola	Fujitsu	Samsung	Infineon
RCA	Philco Ford	NEC	Philips	TI	Renesas
Westinghouse	Philips	Fairchild	Intel	IBM	TSMC
Motorola	Transitron	RCA	NSC	Mitsubishi	Sony
Clevite	Raytheon	GE	Matsushita	Philips	Philips
$75	$1,528	$4,012	$21,479	$144,404	$227,484
Discretes	Std	Logic	MOS	CMOS	MPU

Dataquest

Will you be a global player?

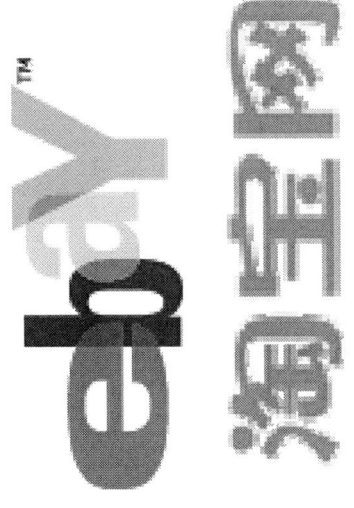

Limited success in China

IP Model enables diversity

Sharp IT-32X2
LCD Television

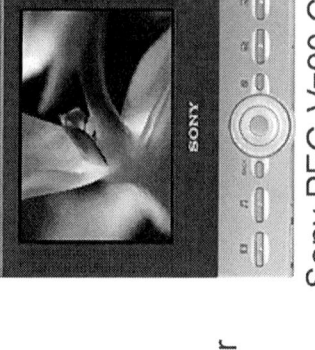
Sony PEG-Vz90 Clie
Personal Digital Assistant

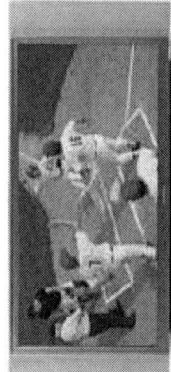
Toshiba Gigabeat
ARM1136J-S™

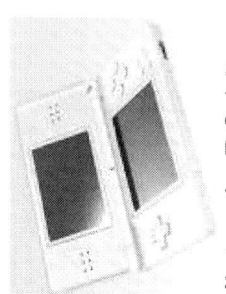
Nintendo DS-Lite
ARM9 & ARM7

Samsung Camcorder
ARM9

Toshiba 52HM84
52" DLP Television

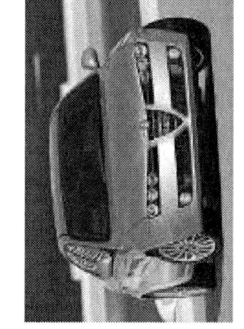
Alfa Romeo Brera &149
ARM9™ & ARM7™

LEGO Mindstorm NXT
ARM7™ family-based MCU

JVC Digital
Camcoder
ARM7TDMI®

Archos AV400
PMP ARM9

Nokia High End Consumer
Phones

Samsung Blu-Ray DVD
ARM926EJ-S™ + JTEK S/W

My Favourite Printer

- Z Corporation ZPrinter® 450
- Output precise, 300 x 450 dpi color
- Layer 0.1 mm. at 2 to 4 layers a minute

Just add consumables.......

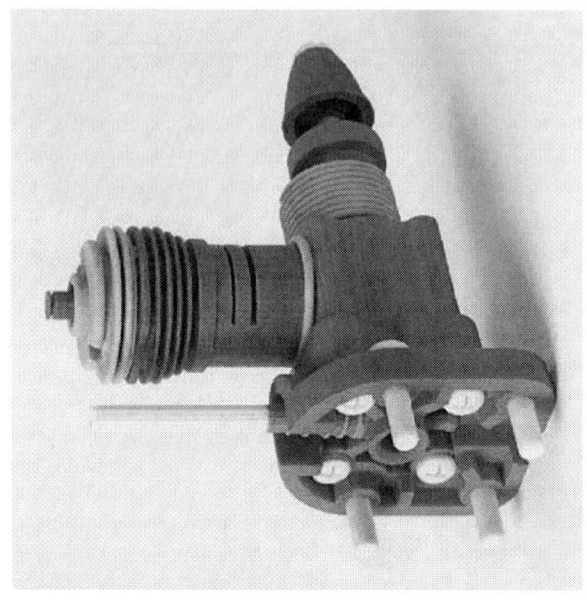

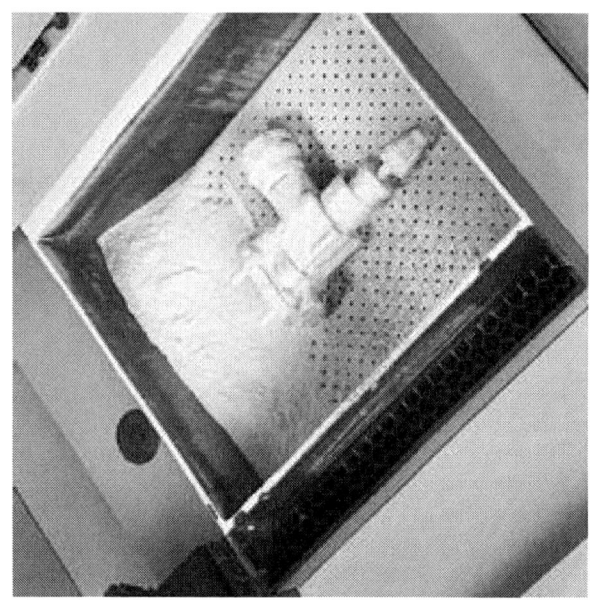

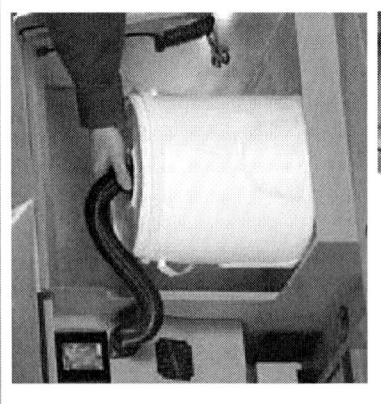
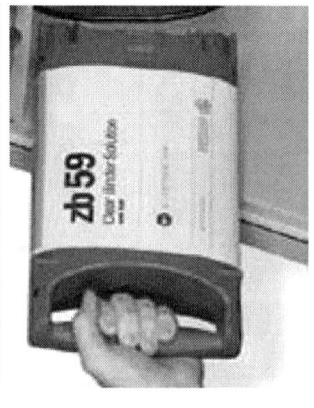
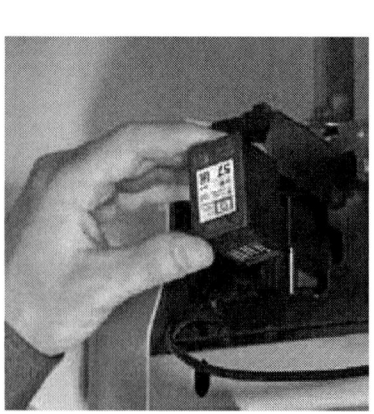

1,600,000,000 ARM Powered® processors in mobile devices in 2006

ARM

at the heart ... of wireless evolution

Telephone industry mature in 80s?

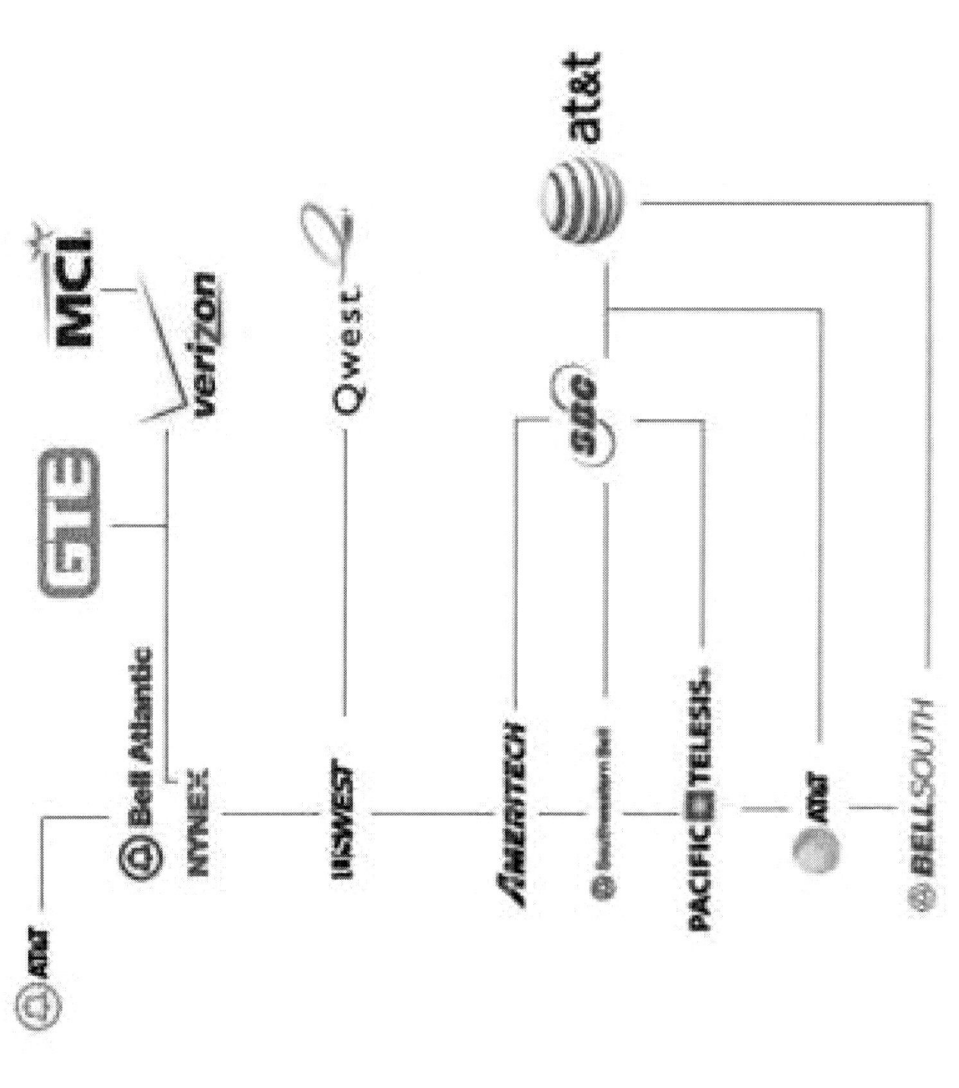

Services Driving Processor Needs

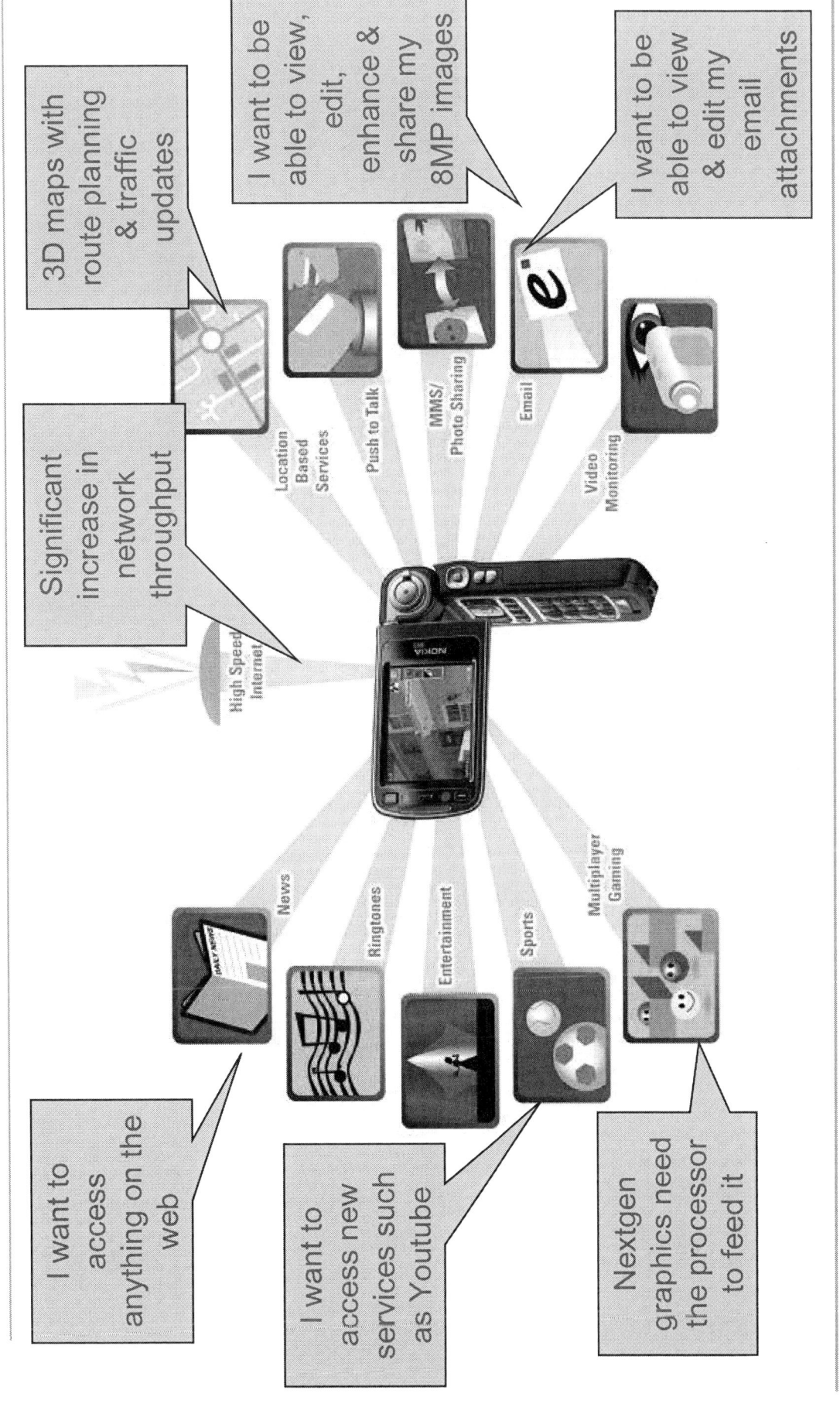

PC style applications move to phones

- Increasing software complexity is driving operators towards Symbian/Windows Mobile/Linux

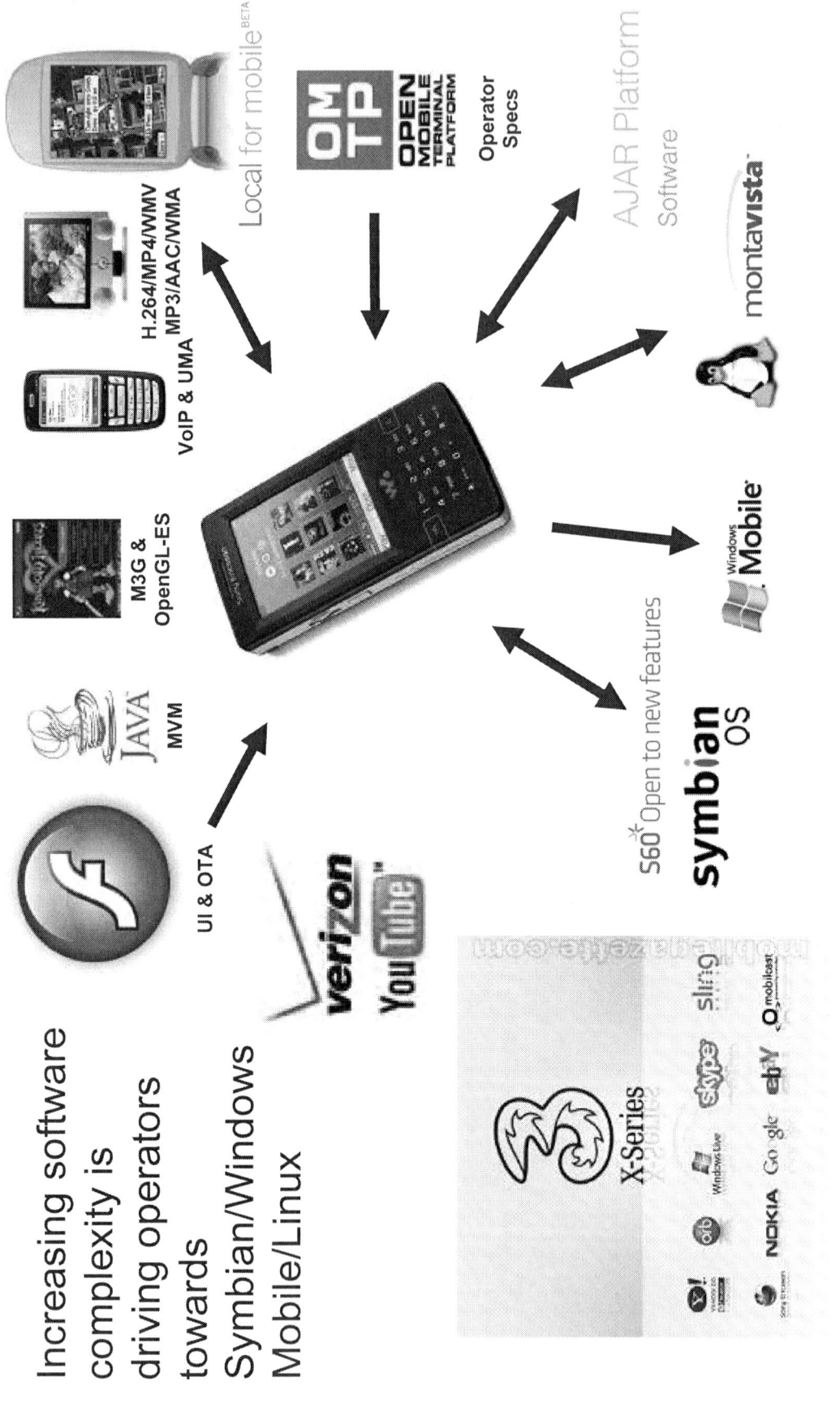

Some things don't change some do....

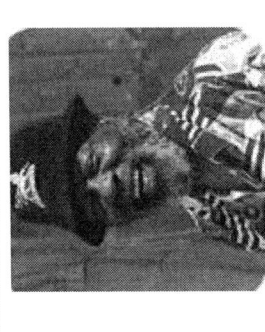

- March 2007 U-Roy plays the Junction
- Who I first saw 30 years earlier......
- Live music is the same
- But I used to buy
  - Vinyl
  - Now I buy CDs
- With the EMI deal I might buy digital

- And I used to go to the box office
  - Now I buying tickets on the internet
  - Or Ebay

Some things change

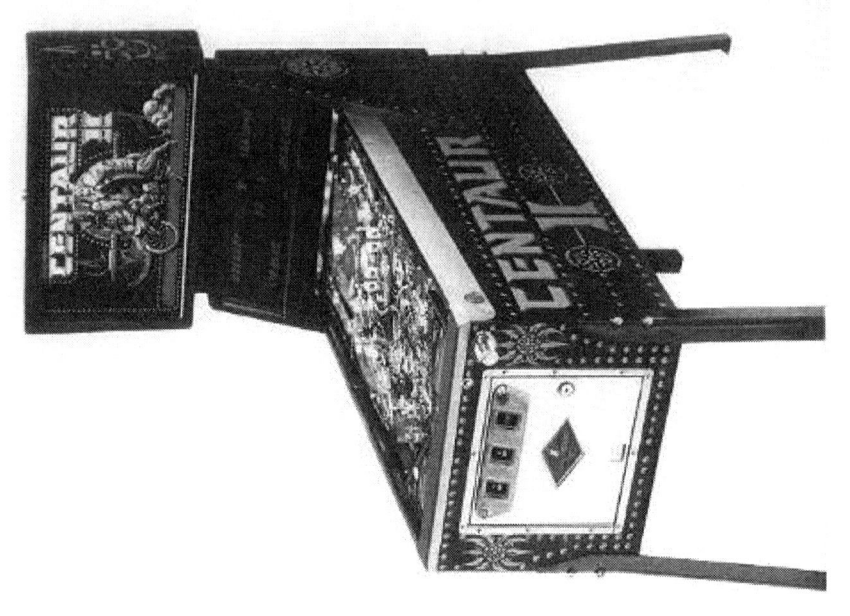

Bally Manufacturing Corporation
(1931-1983)

8 bit controller and 2 flops to a chip

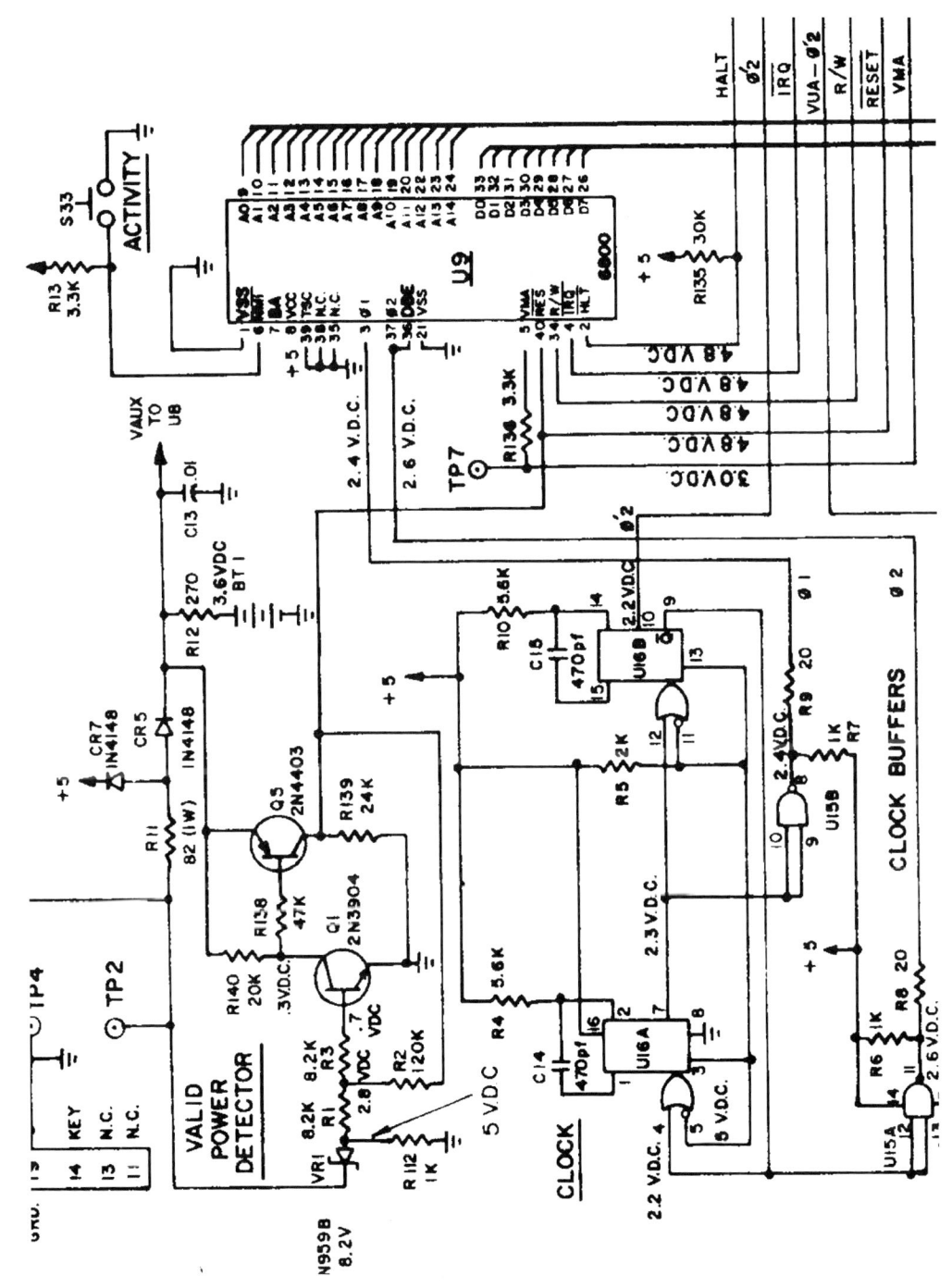

System Architecture Problems Today

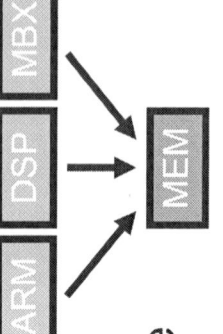

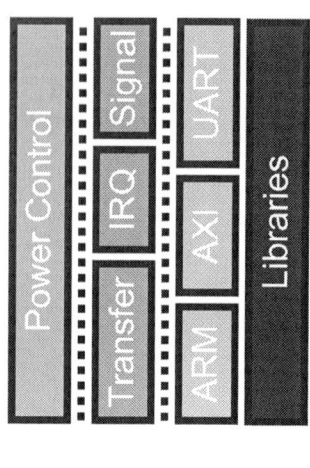

- Resource Sharing (horizontal, many to one)
  - Multiple IP blocks come together on a SoC
  - IP blocks are shared e.g. Memory, Busses, DSP, Core
  - Protection, Security, QoS

- Application Partitioning (horizontal, one to many)
  - Applications require distributed resources
  - Core + DSP, Multi-processing, Accelerators
  - Signalling, Communication, Single-Source Compiler

- System Layering (vertical, many to many)
  - Enable Differentiation and "Technology Towers"
  - Portability, Compatibility, Reliability
  - Start-up Configuration, Discovery, PM

Everything is getting harder

We can't…

Design it right — Increasing complexity - Comprehensive simulation and testing of a design becoming more of an issue

Build it right — Atomic variability – different errors on a die by die basis; Contamination during fabrication – DFM

Keep it right — Significant increase in SEUs in logic due to reduced stored charge; Increasing wear out mechanisms exacerbated by shrinking geometries (electromigration, oxide breakdown, hot-carrier interaction)

Test it right — Increasing testing times; Correlation between single defects and multiple failures reducing; How do we test for susceptibility to intermittent defects?

Size, Power, MHz the ride is over

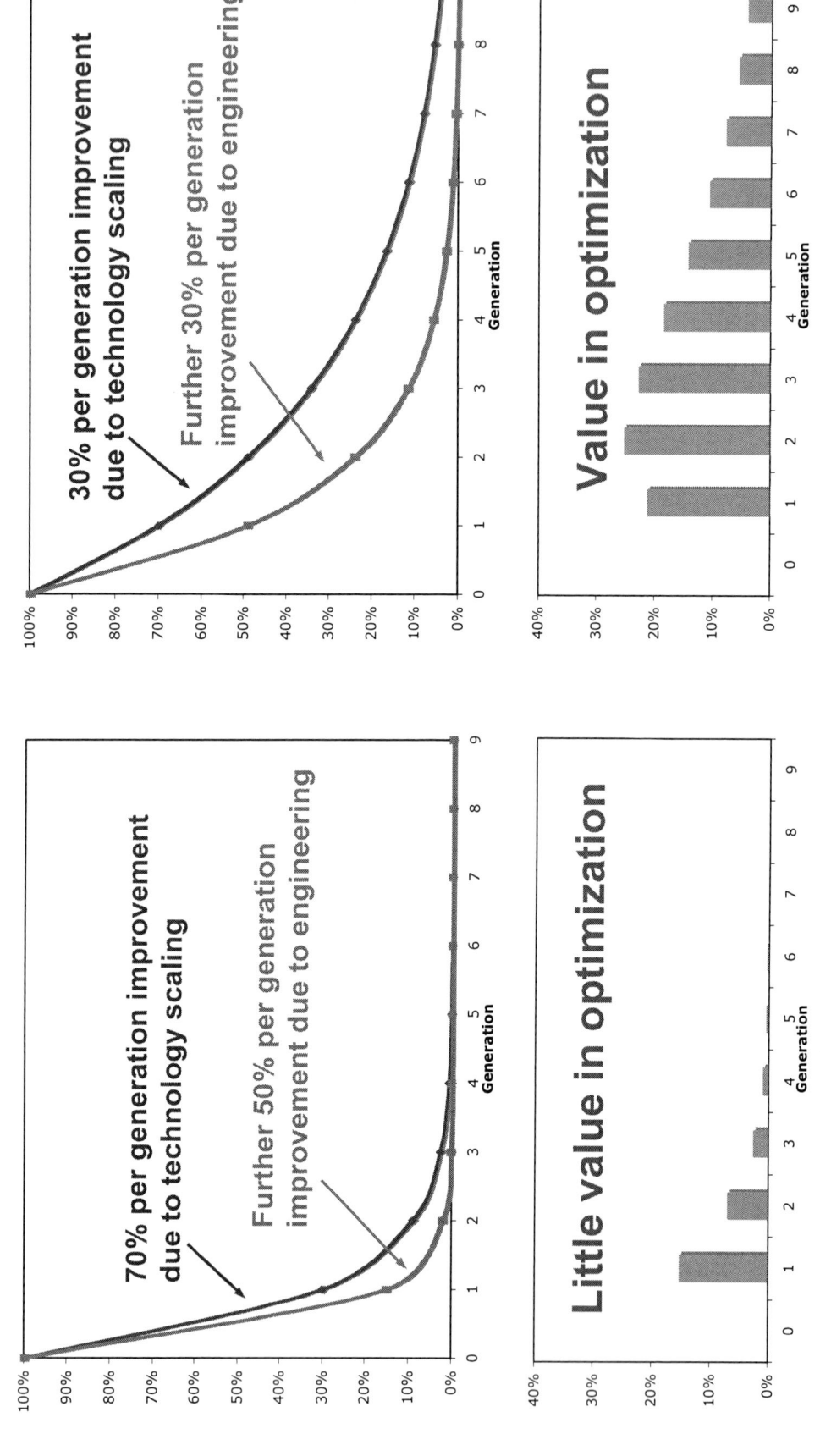

Performance/Power Flexibility

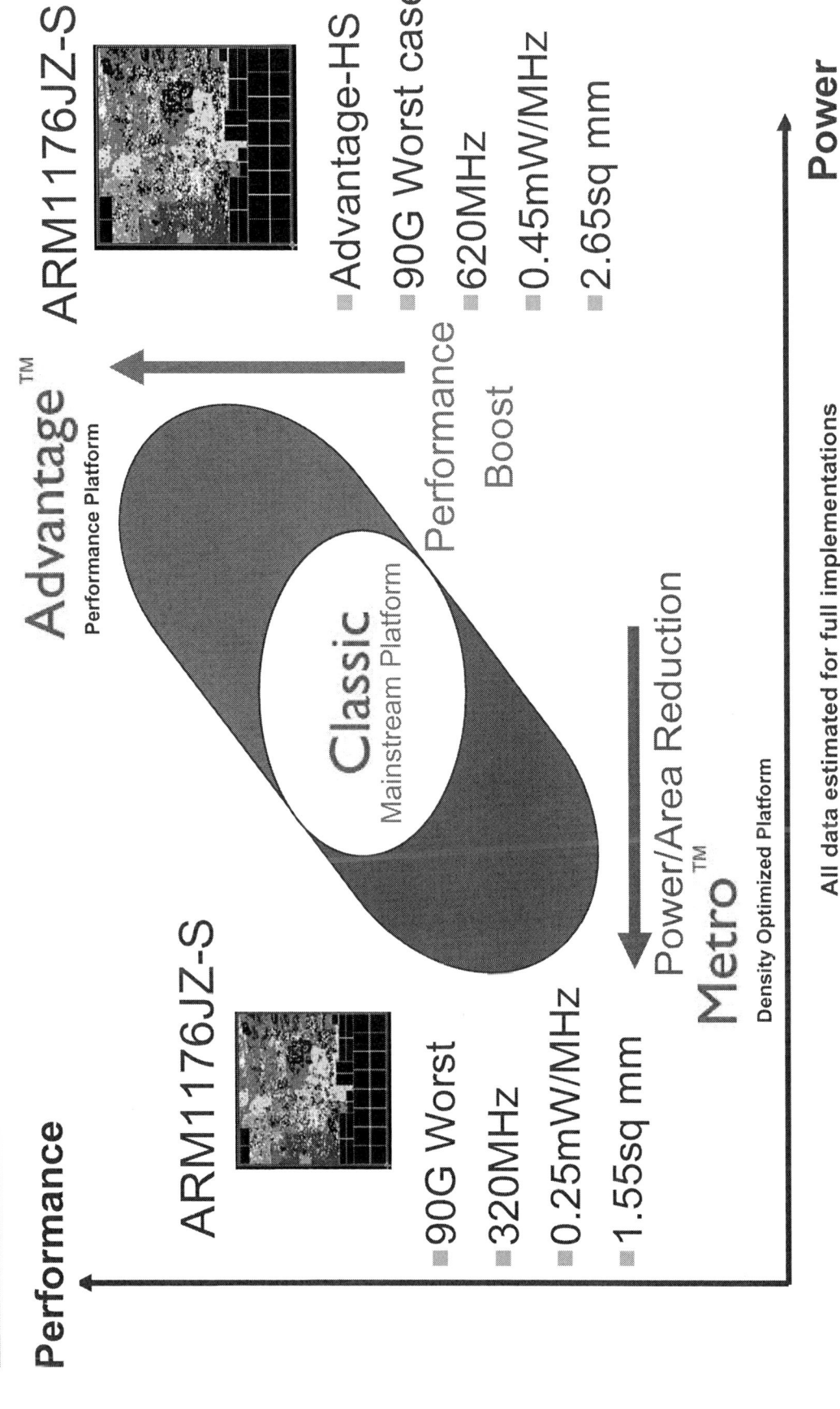

Performance

Advantage™
Performance Platform

ARM1176JZ-S

- Advantage-HS
- 90G Worst case
- 620MHz
- 0.45mW/MHz
- 2.65sq mm

Performance
Boost

Classic
Mainstream Platform

Power/Area Reduction

Metro™
Density Optimized Platform

ARM1176JZ-S

- 90G Worst
- 320MHz
- 0.25mW/MHz
- 1.55sq mm

Power

All data estimated for full implementations

Software, Key to SoC Design

250-500K Lines of F/W

50-100K Lines of Protocol F/W

TV Decode

Over 2M Lines of Application S/W

250-300K Lines of DSP F/W

xDSL

Up to 2M Lines of Network S/W

5-10K Lines of Control Code

Video Display

>100K Lines of Appl S/W

20-50K Lines of Protocol F/W

Wireless

5-10K Lines of Microcode

Nearly 5 Million Lines of Code necessary to Enable such a Chip

NAND Flash Demand Drivers

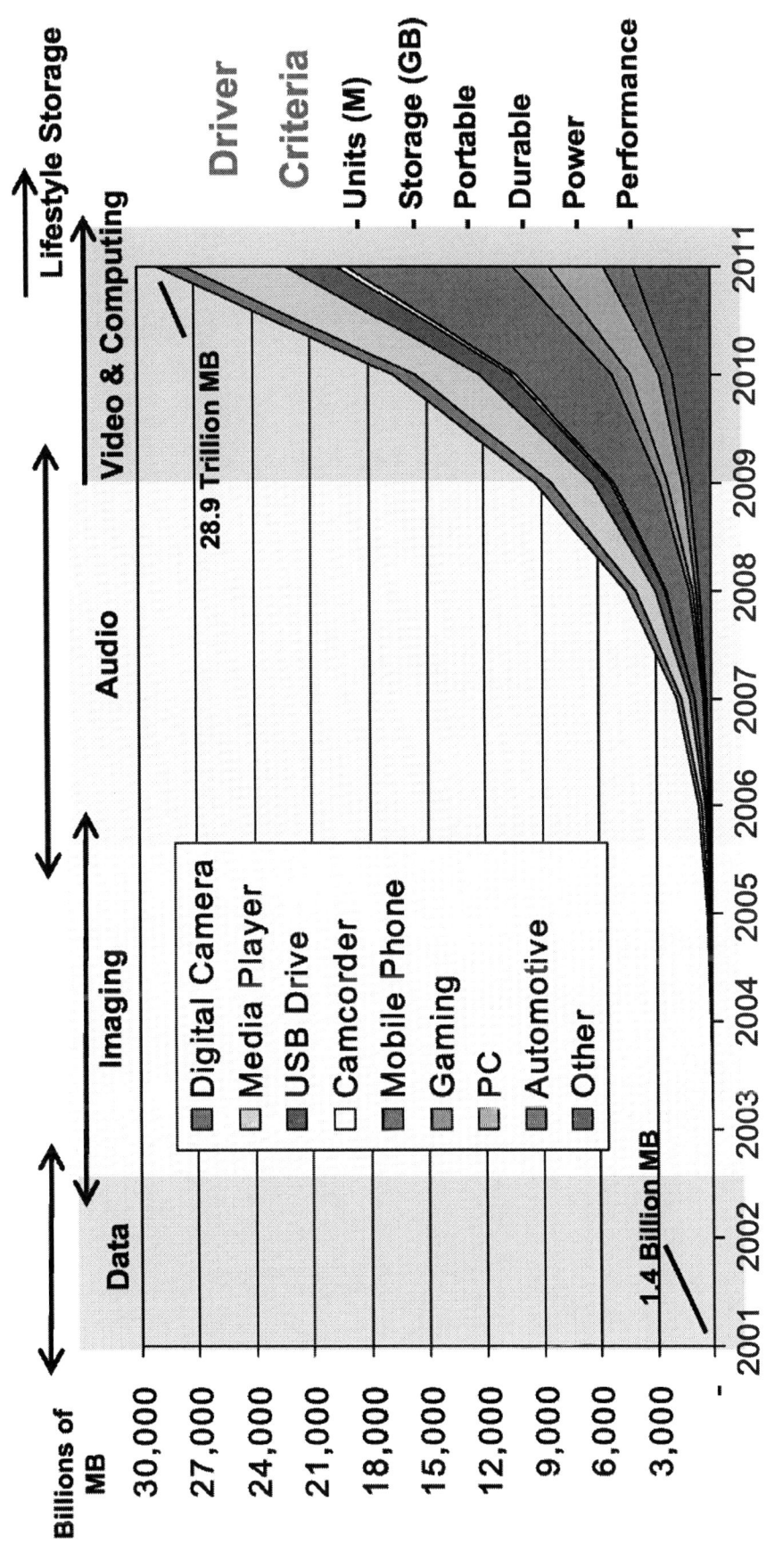

How do you build heterogeneous MP

Design
- The right system, with the right components, efficiency, flexibility, and performance

Implementation
- Within acceptable energy and yield envelopes
- Without breaking the verification budget

Programming
- To preserve the software investment
- Without overly constraining platform choice

Process technology has a significant impact on the choice of optimal system design

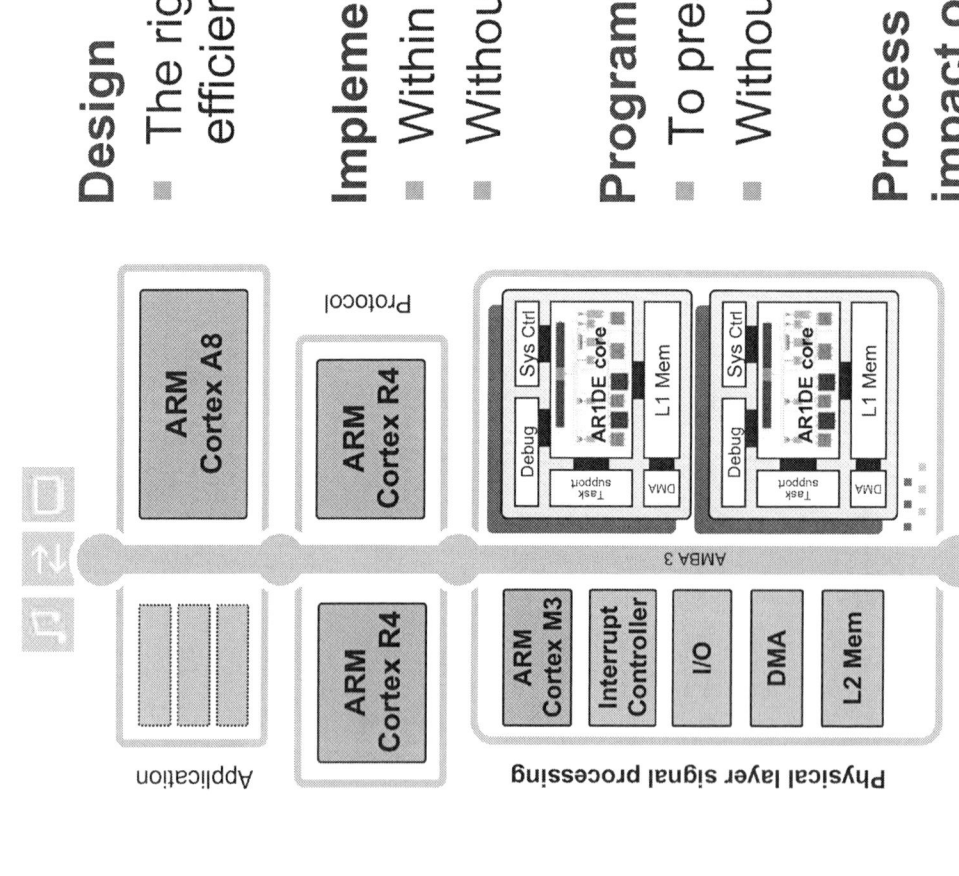

The AR1DE core

Built using OptimoDE technology

512 bit wide SIMD

- 32 x 16 bit concurrent operations
- Block floating point support
- 15 x 512 bit registers in SIMD register file
- 15 x 32 bit registers in SIMD predicate register file

Permutation function unit

- Good match for key kernels: FFT, Viterbi, ...
- Both forward and backward configurations reduce number of iterations for arbitrary permutations

Wide superscalar execution

- Concurrent SIMD, scalar, and memory operations

Fully compiler-exposed microarchitecture

- Leads to efficient hardware implementation
- Compiler scheduled operations reduce hardware complexity

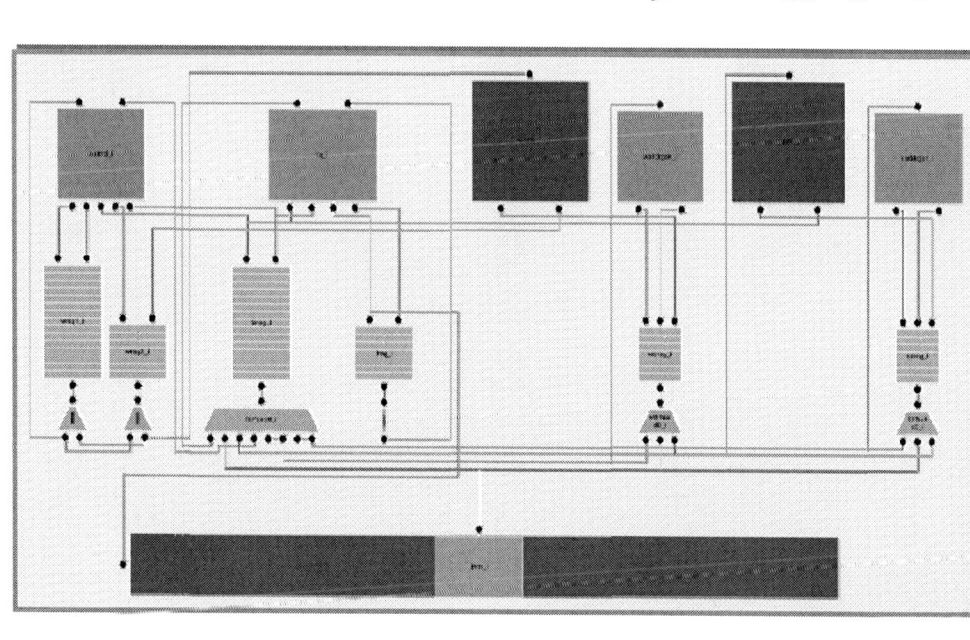

The cost of binary compatibility

- For ARM11 integer pipeline control logic is:
  - 40%-45% of the total area
  - 50%-55% of the total power
  - Timing critical paths are in the control logic

- Estimated cost for adding binary compatibility:
  - 30%-40% increase in area and power
  - 10%-20% drop in frequency

- Portable source enables further efficiency gains in the future

- Evolution is not a straight line: different intrinsic operations make sense for different workloads and technology points

Programming a cluster

Two level application description

- **System-level**: describes inter-kernel communications, timing constraints, mapping attributes, assertions

  Concurrent tasks extracted from "C + channels + attributes" description

- **Kernel-level**: "C + primitives + vectors"

System compilation

- Generates tasks, schedules, communication stubs, DMA requests, timing assertions, synchronization, debug support

- Iterative compilation: marks up program with increasingly detailed attributes; fully attributed program required for final mapping

Kernel compilation

- Standard "C + primitives + vectors" converted to machine code

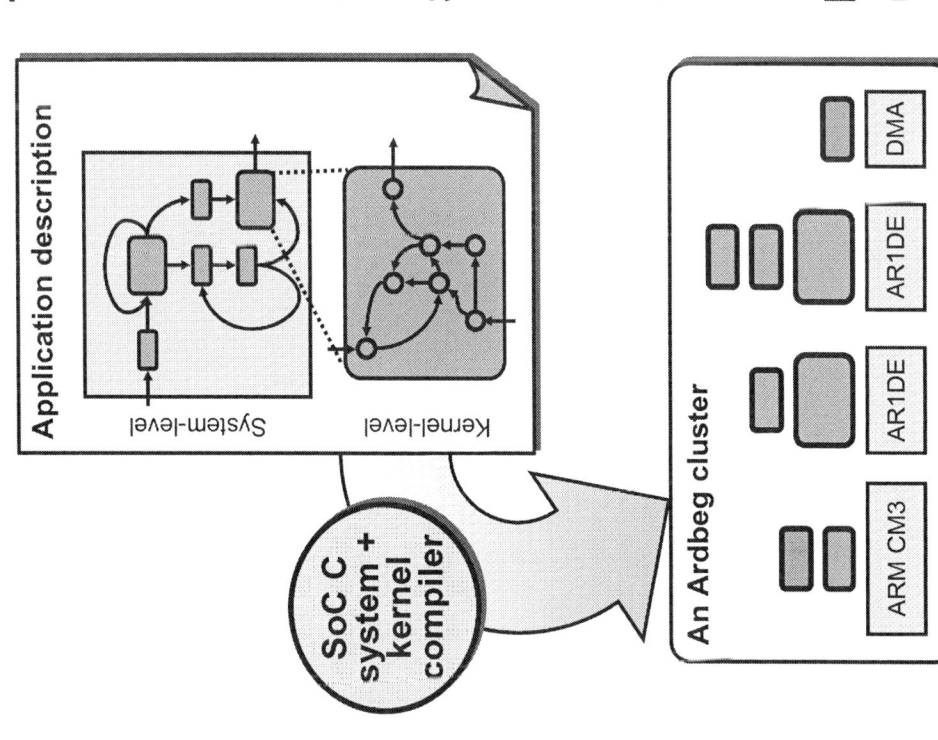

Massively Parallel Systems

- Neurons A flexible biological control component
  - Humans: 10^{12}
  - Bees: 850,000
- Not like a CMOS circuit or a CPU
  - have fan-in/fan-out of ~1,000
  - low-performance components ~100 Hz
  - low-speed communication ~metres/sec
- So lets use CMOS and CPU's to simulate
  - Goal 10^9 neurons
- Spinnaker Project

Neural Maths (very approximate)

- 10^9 neurons — with 1000 fan in/out — 10^{12} synapses
- 10^{12} synapses — with 1 Bytes per link — 10^6 Mbytes
- 10^{12} synapses — mean switch 10Hz — 10^{13} links/sec
  - It's all about the network
- 10^{13} links/sec — with 10 instruction/link — 10^8 MIPS
- 10^8 MIPS — with 100MHz ARM — 10^6 ARMs

- 1 ARM — with 100MHz ARM — 1000 neurons
- 1 Chip — 20 ARM968 + 128MB — 20K neurons
- 50K Chips — with 20K neurons — 10^9 neurons

- 25uW per neuron for complete CPU + Links

50 years of progress

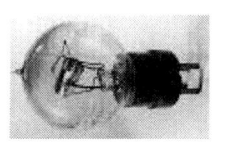
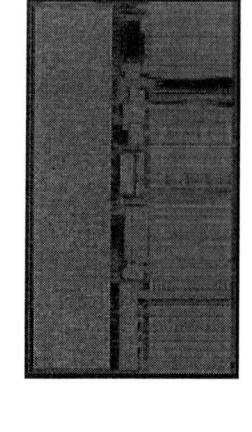
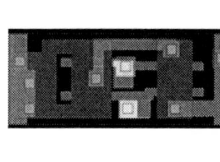

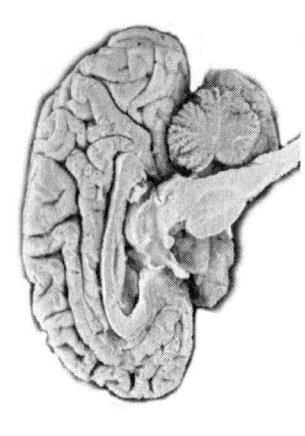

- Baby
  - filled a medium-sized room
  - used 3.5 kW of electrical power at 700 Hz
- ARM7TDMI
  - 0,1mm2 with 100K Transistors
  - uses 9mW at 165 DMIPS
  - 10^{10} times better power consumption
- Spinnaker
  - 25uW per neuron for complete CPU + Links
- You and me
  - 25pW per neuron
  - 10^6 power efficiency to find……

A CMOS replacement?

- Carbon nanotubes are among the numerous candidates for tissue engineering scaffolds since they are biocompatible, resistant to biodegradation and can be functionalized with biomolecules.

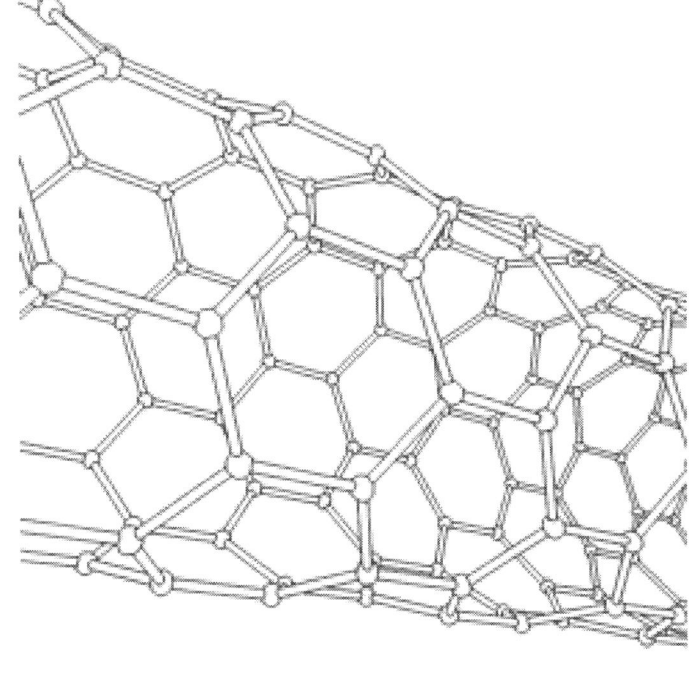

Effective Chip Packaging
.... starts with Design

Andy Longford
Technical Consultant, PandA Europe, UK.

PandA Europe
Chip Packaging and Assembly Consultants

The Final Manufacturing Processes –

Yesterday .....

- Receive tested wafer
- Dice
- Wirebond
- Mold
- Trim Form Singulate
- Mark
- Test
- Ship

Tomorrow .....

- Manufacture wafer
- Test
- Ship

...... why ?

FSA Paris May 2007

Packaging Aspects

High performance chips must consider effects of Packaging:

- Stresses
- Strains
- Parametric Discontinuities
  - Interconnect path lengths
  - Inductances
  - Capacitances
  - Resistances   etc ...

FSA Paris May 2007

SiP is the (ITRS) Future

Chip / Component Configuration	Technology
Side by Side	Substrate:

Table 96a   Chip to Package Substrate—Near-term Years

Year of Production	2005	2006	2007	2008	2009	2010	2011	2012	2013
DRAM ½ Pitch (nm) (contacted)	80	70	65	57	50	45	40	36	32
MPU/ASIC Metal 1 (M1) ½ Pitch (nm) (contacted)	90	78	68	59	52	45	40	36	32
MPU Physical Gate Length (nm)	32	28	25	22	20	18	16	14	13
Wire bond pitch—single in-line (micron)	35	35	30	30	25	25	25	25	20
2-row staggered pitch (micron)	45	40	35	35	35	35	35	35	35
Three tier pitch pitch (micron)	45	40	35	35	35	35	35	35	35
Wire bond—wedge pitch (micron)	30	25	25	25	20	20	20	20	20
Flying lead pitch (micron)	35	35	35	35	35	35	35	35	35
Flip chip area array pitch (micron)	150	130	120	110	100	90	90	90	90
Flip chip on tape or film pitch (micron)	35	30	30	25	25	20	20	20	20

stacked functional layers

FSA Paris May 2007

Development of Advanced Packaging

Wafer & *PWB technology*

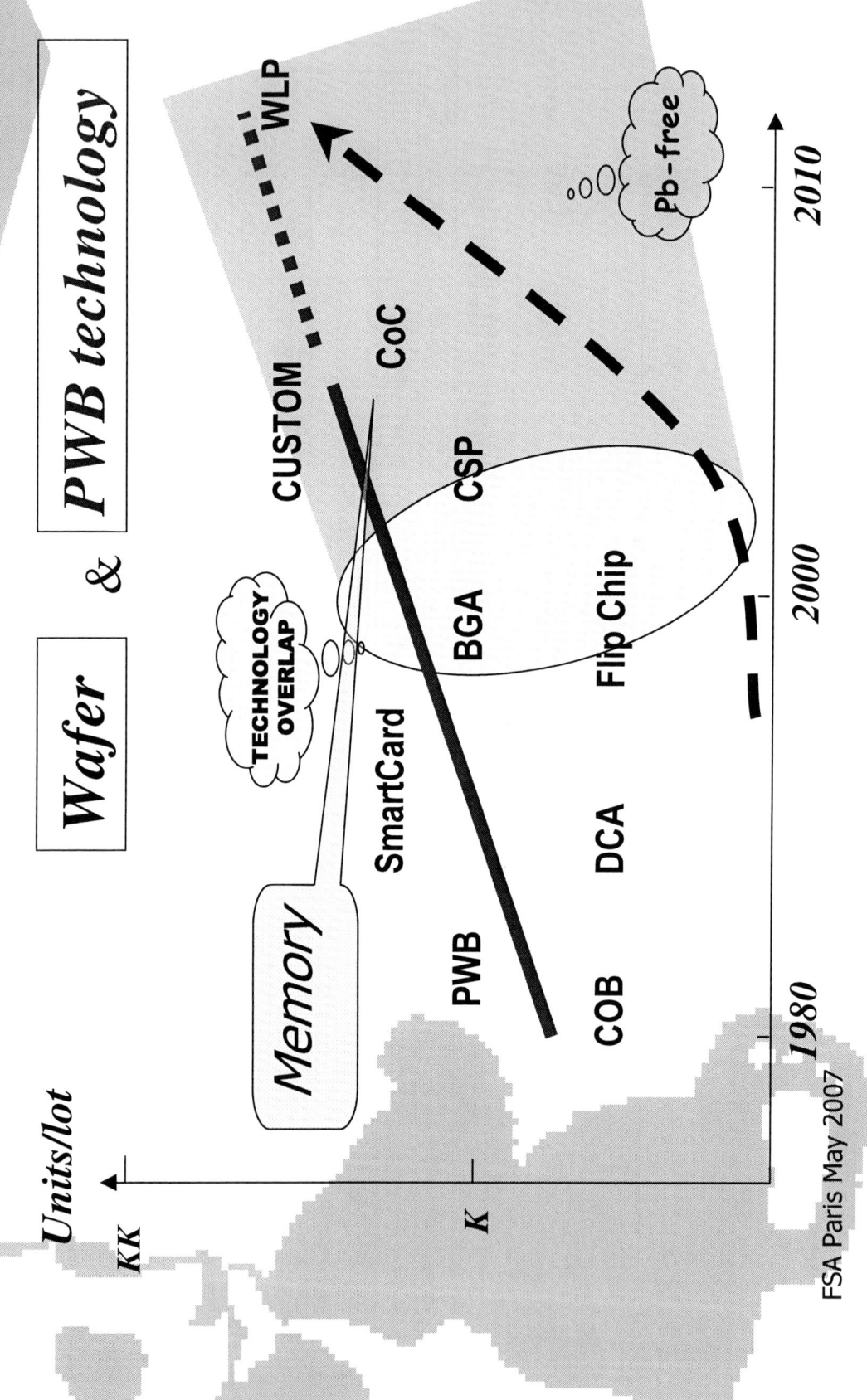

FSA Paris May 2007

Design For Manufacture

In the application of an "Effective" process for Chip manufacture, there needs to be 3 interleaved aspects:

A) Design and manufacture of the components to suit

B) ....the design and manufacture of a package technology that will ....

C) ....be assembled and tested in a volume process.

The parts are the 3 points of a triangle each interconnecting with one another

A

B

C

DXT - DiT

DXP - DiP

DXM - DiM

FSA Paris May 2007

Package Technology Drivers

- ITRS Roadmaps ...
- Market Pull or "envelope" push
- Effect of package parameters ...
- Fabrication Infrastructure (design support?)
- Time to Markets
- Ways to lower cost

..... The Grand Challenge is ?

Cost ......and who pays ?

FSA Paris May 2007

RF SiP

RF System in Package - Technology Choice and Design Methodology

Chris Barratt

Insight SiP

Sophia Antipolis France

SiP Panel Session

ROLE OF SiP IN LIFE CYCLE

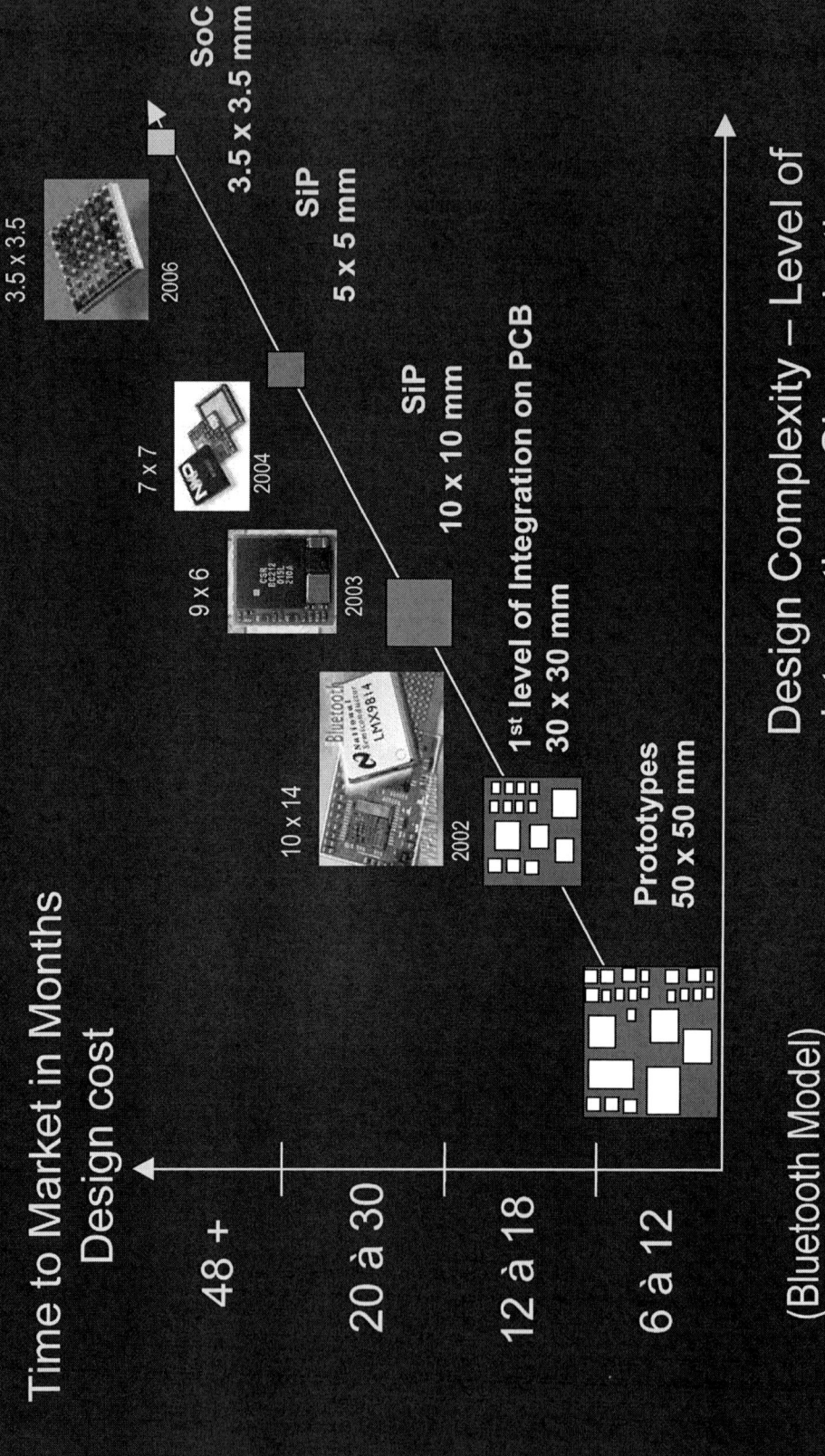

Time to Market in Months
Design cost

48 +

20 à 30

12 à 18

6 à 12

(Bluetooth Model)

Prototypes
50 x 50 mm

1st level of Integration on PCB
30 x 30 mm

SiP
10 x 10 mm

SiP
5 x 5 mm

SoC
3.5 x 3.5 mm

10 x 14
2002

9 x 6
2003

7 x 7
2004

3.5 x 3.5
2006

Design Complexity – Level of
integration – Size reduction

RF SiP Substrate Technologies

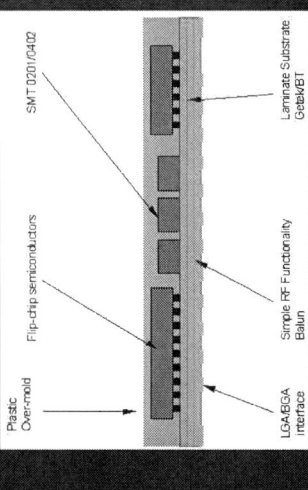 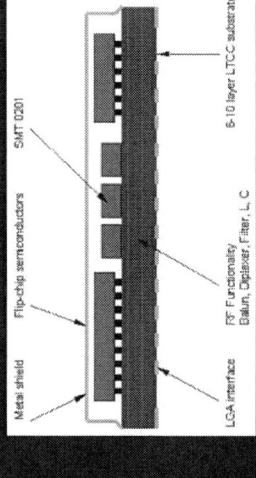 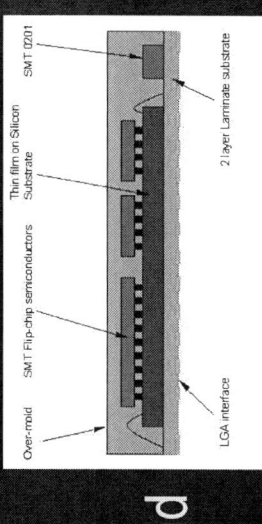

- Laminate based (ε_r 3 – 5)
  - 4 to 6 metal layers
  - Interco. RF Baluns in substrate

- LTCC based (ε_r 7 – 10)
  - 6 to 20 metal layers
  - Interco. RF baluns, Filters, matching in substrate

- Silicon Based (ε_r 11.2)
  - 2 to 4 metal layers + doping
  - Interco. RF baluns, Filters, matching and high C in substrate

RF SiP Design – Choose the best technology

- System Partition
- CMOS, SiGe, AsGa
- LTCC, Laminate, Thin film
- SMT or buried passives
- Wire-bond vs Flipchip
- ➤ Size, Time to mkt, Costs
- Compare Options
- Choose best compromise

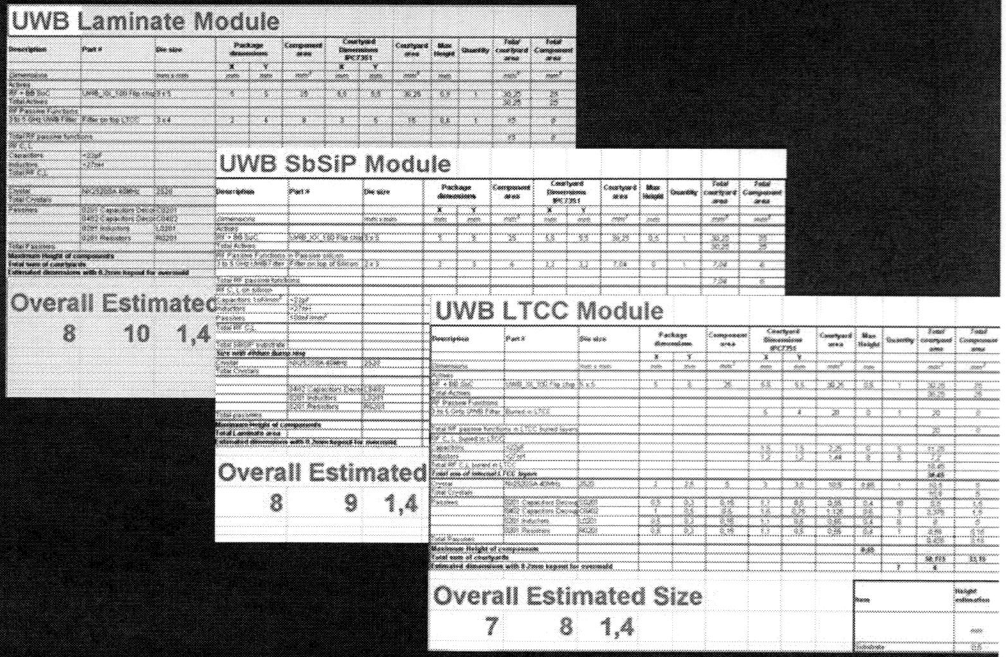

Detailed Design

- Standard EDA software based.
- Flexible process with no fixed libraries for each substrate
- Step-by-step Process
  - System model development
  - Project-specific component library
  - Circuit Optimization taking interactions into account
  - Exhaustive simulation of electro-magnetic behavior
  - Performance Optimization through electro-magnetic simulation feedback to circuit model.
  - Proven two-pass success (with one-pass objective)

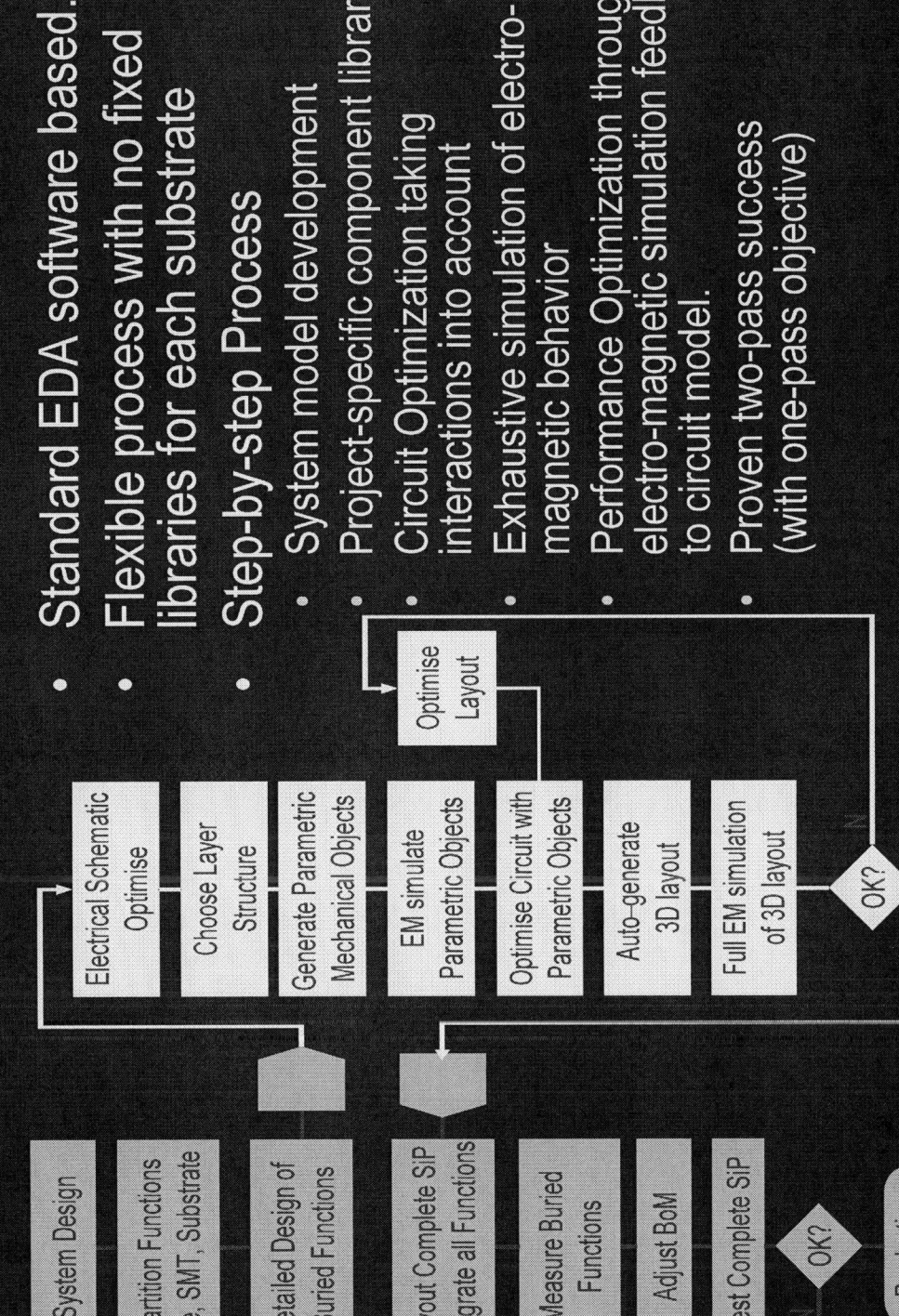

imec

Technology Aware Design

- Technology aware design solves the challenges imposed by scaling (variability, reliability, static power).
- It is complementary to DFM.
- Brings back that "happy scaling feel"

Author: TAD team
Presented by Bart Dierickx
Director of Technology Aware Design program
IMEC, Leuven, Belgium
FSA-IET 14 may 2007

Outline

- Scaling hits fundamental limits

- Strategic "what-if" questions
  - Example 1: Technology options
  - Example 2: Next node opportunities

- IMEC *Technology Aware Design* approach
  - Variability aware modeling – answers what-if questions
  - Runtime countermeasures

- Comparison with state of the art
  - DFM – DFY – SSTA

- Conclusions

imec confidential 2007

Scaling hits fundamental limits

picture: Asenov – TAD workshop 2005

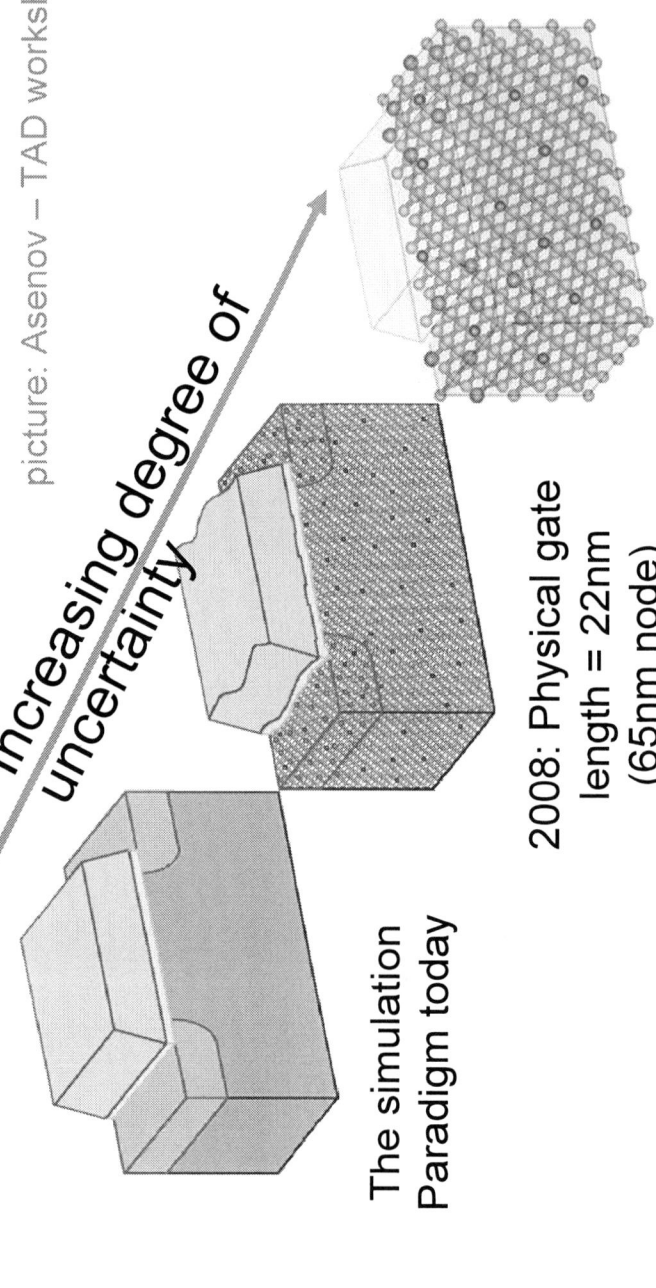

Increasing degree of uncertainty

The simulation Paradigm today

2008: Physical gate length = 22nm (65nm node)

2016: Physical gate length = 9nm = 30x30x30 atoms (22nm node)

➤ Atomic uncertainty in dimension and dopants

Before that other brick walls will have stopped us:

➤ Litho = molecular randomness, wavelenght...

➤ Electric field, strain/stress, leakage

➤ Economical

imec confidential 2007

Physical gap: 5 issues

- In the happy scaling era speed, cost and power budget were the only constraints; design and technology did not interact

Scaling induced new awkward constraints

  – Variability

  – Leakage, static power dissipation

  – Reliability, finite and unpredictable lifetime

  – Interconnect scaling issue

Some issues have found separate SotA solutions

  – Solved in the technology itself (high K, backend, novel device structures...).

  – Ad hoc design solutions (footer switches, guard banding, multi VTH, low VDD...)

5 persistent issues, requiring design-technology interaction

  – 1. system-level yield, strongly affected by variability and reliability

  – 2. timing-/energy-closure in the presence of variability

  – 3. interconnect scaling impact

  – 4. dealing with new devices

  – 5. system reliability/life-time

Industry hits problems that can not be solved standalone

Speech bubbles:

- Can I reach my power budget?
- How does low-k affect my yield?
- Can I further reduce VDD?
- Must I fear ...bility?
- Please do not change design paradigms!
- Field returns?
- ...libraries?
- Redundancy at all levels?
- New thoughts on ...?
- Corrections
- FinFETs?
- Is ultra low VDD a solution?
- Must I pay for process options?
- Maybe high-k is a mistake?
- ...is variability the limit or reliability?
- Yield impact of process option?

"what are the next node design issues".
- Predicting and modeling the imperfection.

"how to handle unreliability"
- Design (circuit, architecture, system) solutions that will allow to live with variability and reliability issues.
- Systems must=will allow that a fraction of the devices is weak or fails, at first or during life time.

"Which technology imperfection *is / is not* allowable"
- Investments in technology options becomes a G$ risk
- Be sure that systems can use it, and predict the consequences

imec confidential 2007

imec

48

Fabless industry
and strategic "what-if" questions

- Example (1) trade off speed vs leakage: technology options versus yield

- Example (2) next node is faster, but has higher variability: yield impact?

Cases are real,
Numbers are simplified

- What does it take to apply these examples to your case?

imec confidential 2007

Tradeoff speed versus leakage technology options

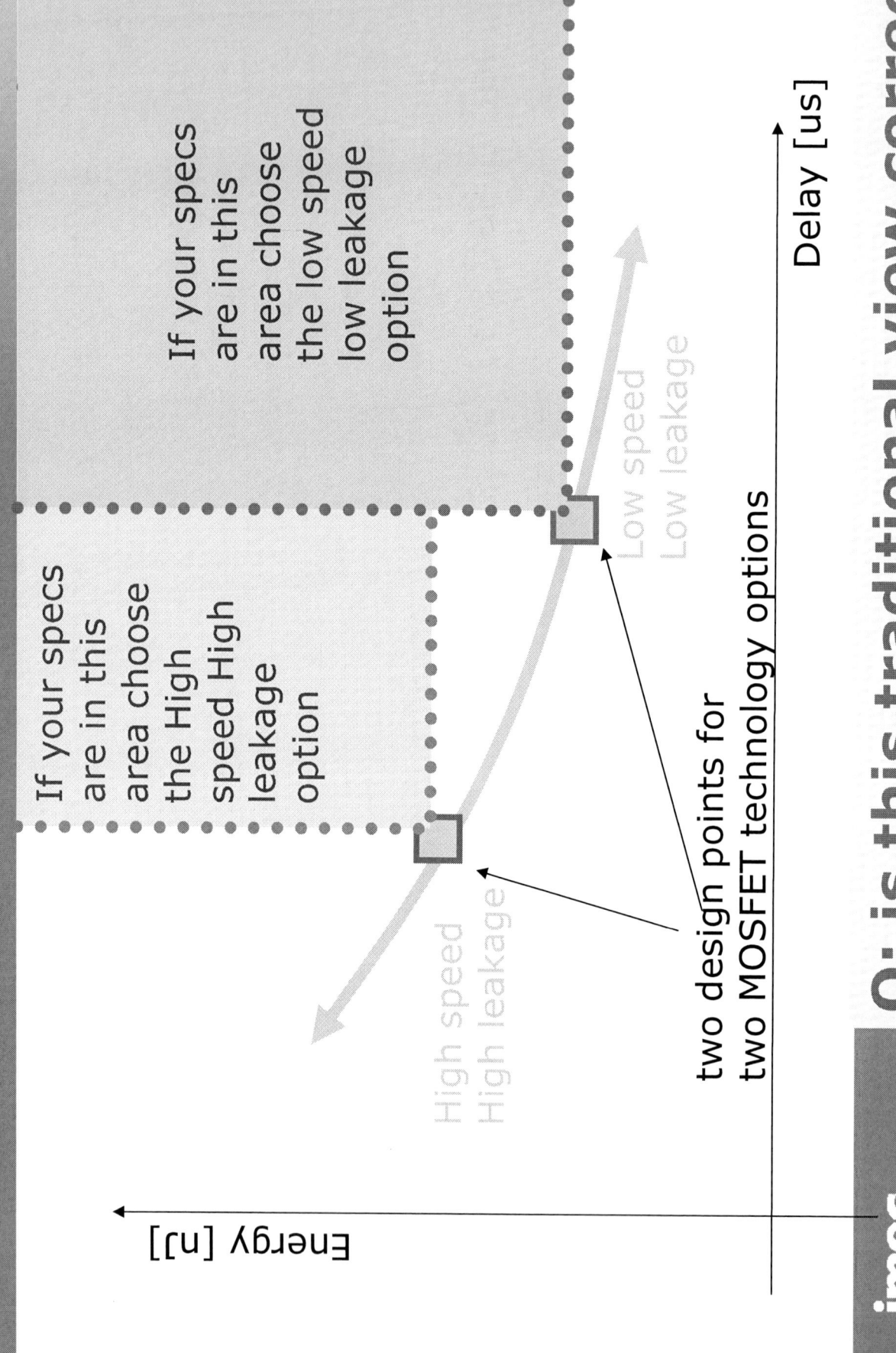

If your specs are in this area choose the High speed High leakage option

If your specs are in this area choose the low speed low leakage option

High speed
High leakage

Low speed
Low leakage

two design points for
two MOSFET technology options

Energy [nJ]

Delay [us]

imec Q: is this traditional view correct?

Tradeoff speed versus leakage technology options

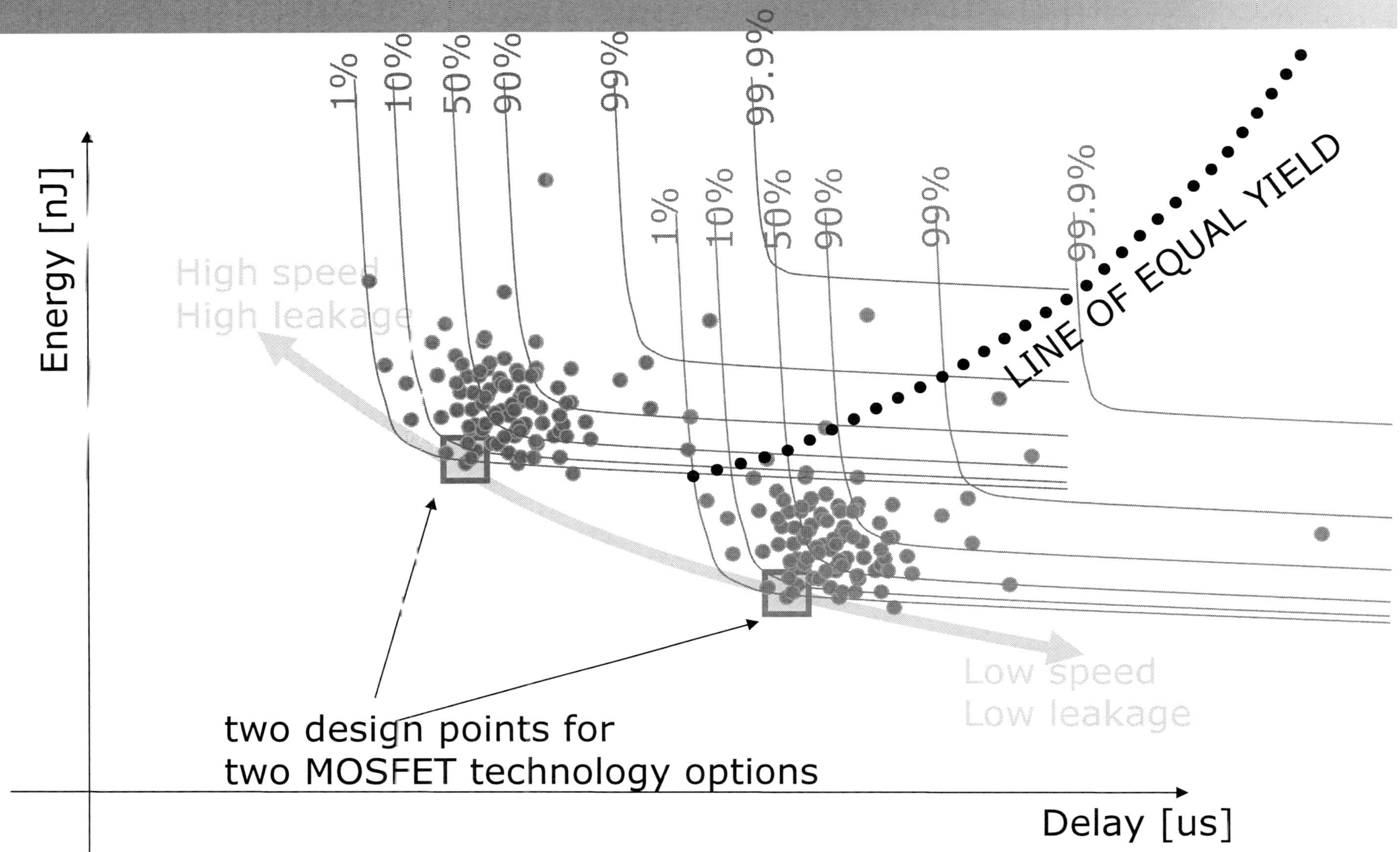

LINE OF EQUAL YIELD

99.9%
99%
90%
50%
99.9%
10%
1%
99%
90%
50%
10%
1%

Low speed
Low leakage

High speed
High leakage

two design points for
two MOSFET technology options

Delay [us]

Energy [nJ]

imec confidential 2007

Tradeoff speed versus leakage technology options

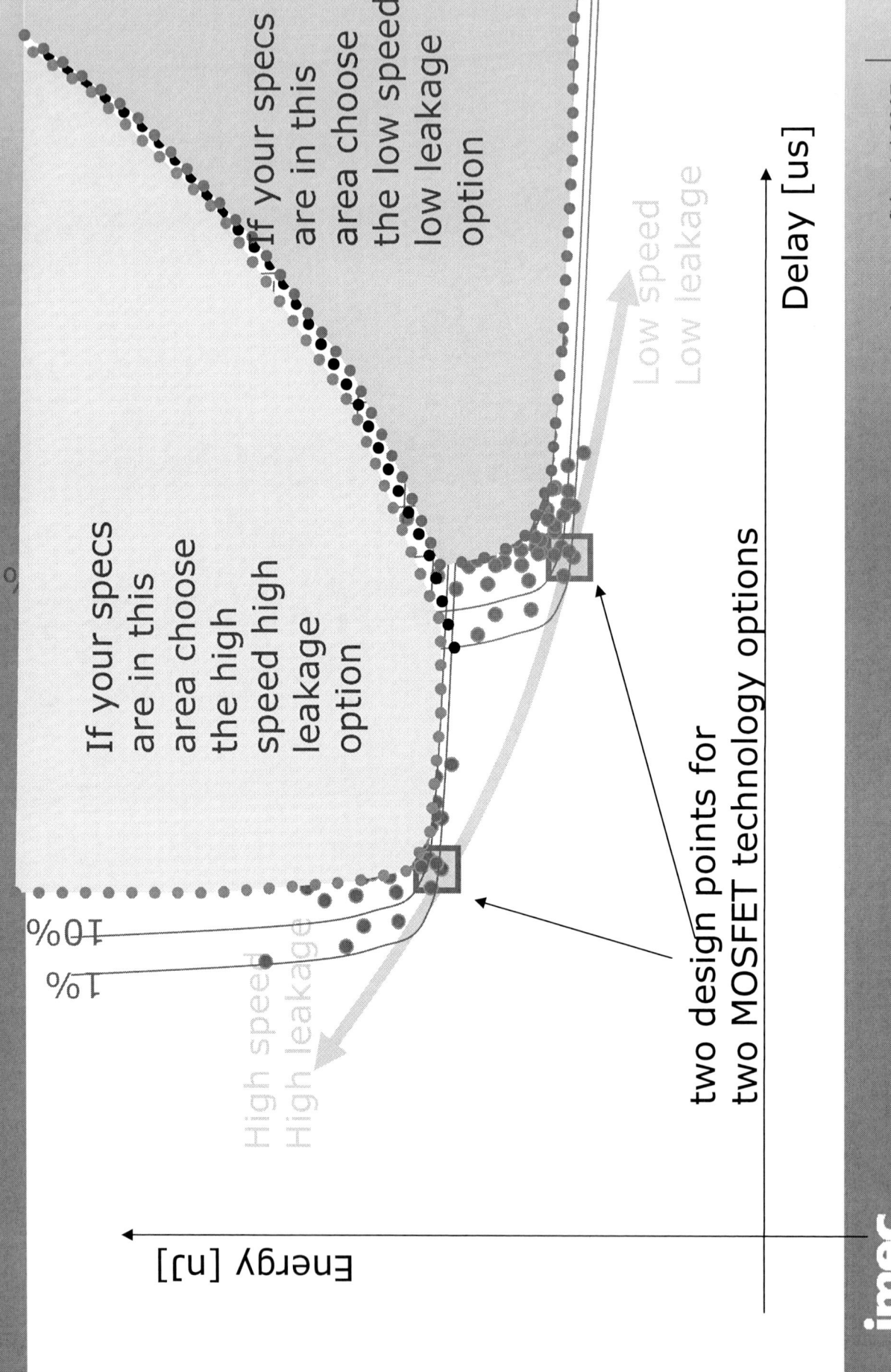

If your specs are in this area choose the high speed high leakage option

If your specs are in this area choose the low speed low leakage option

Low speed
Low leakage

High speed
High leakage

two design points for two MOSFET technology options

10%

1%

Energy [nJ]

Delay [us]

imec confidential 2007

Next node is faster but has higher variability

Similar issue: high-K versus low-K

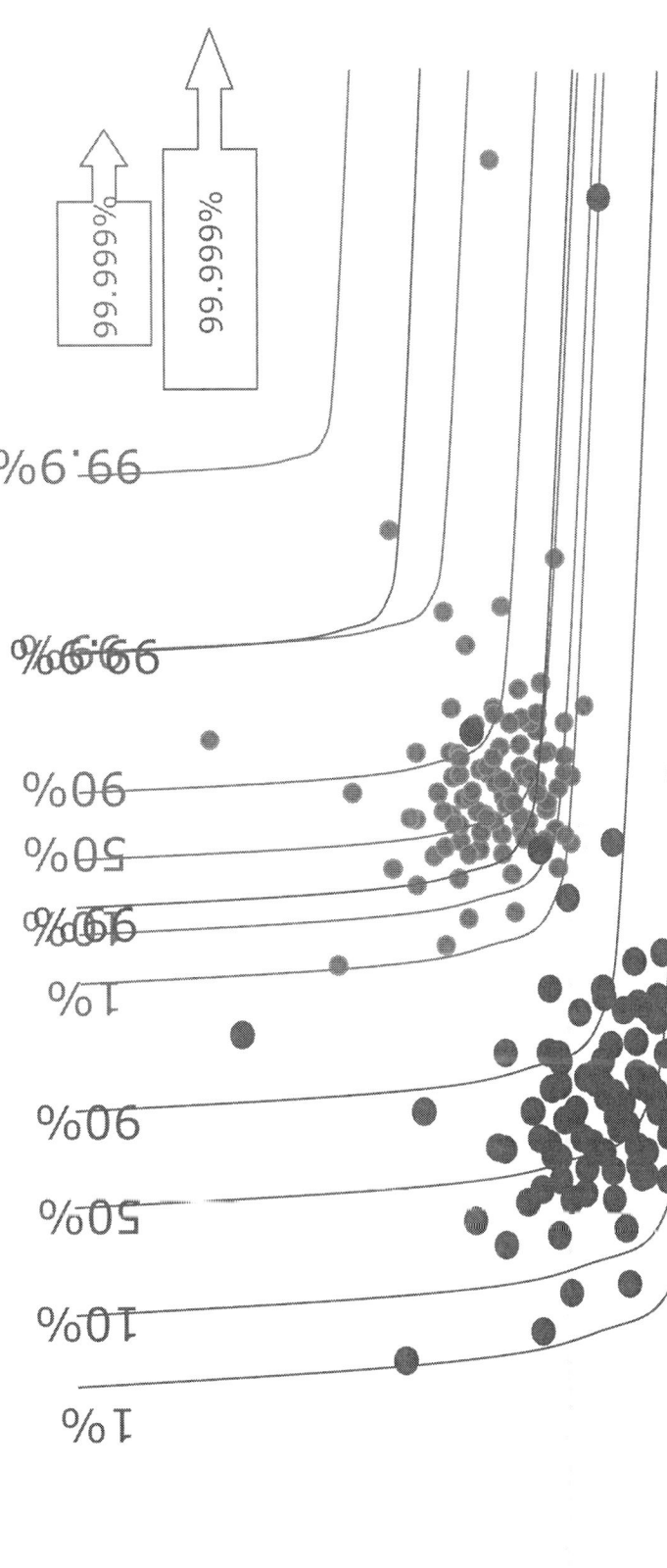

Delay [us]

Energy [nJ]

imec confidential 2007

Next node is faster but has higher variability (2)

The poison is in the tail

Parametric yield
due to delay
depends on the
worst transistors
or gates

Next node:
On the average faster,
but wider distribution...?

log(# samples)

Delay [us]

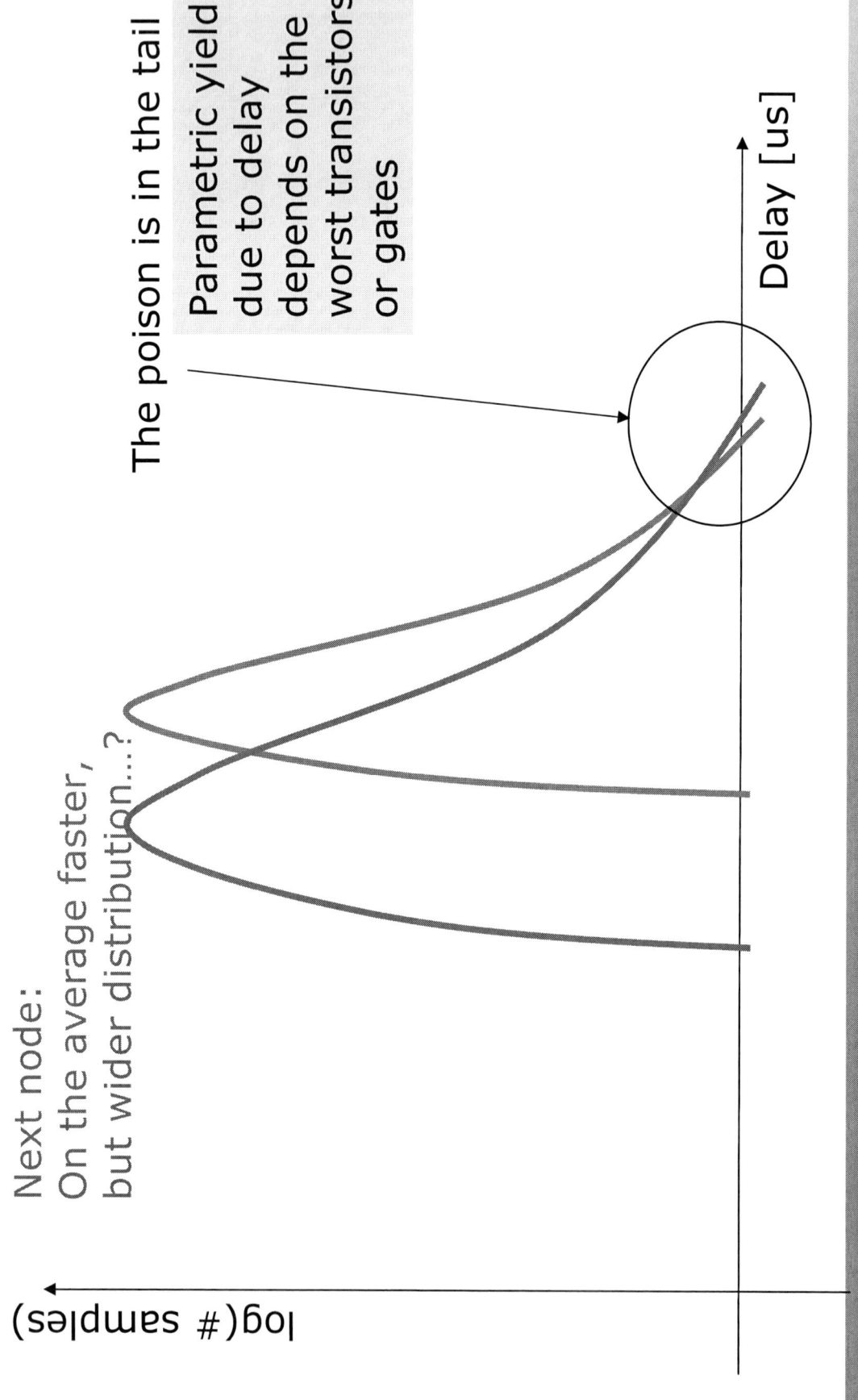

imec confidential 2007

What is "Technology Aware Design"

IMEC's Technology Aware Design program

- physical design solutions for DDSM scaling issues, needing know-how bridging technology and design

- Focus on analysis (VAM) and solutions (SKM)

- www.imec.be/tad

Focus 1 → **ANALYSIS**

VAM: Variability & Reliability Aware Modeling
- Understanding, modeling, predicting, predicting the unknown
- Technology – System yield coupling
- Strategic what-if questions
- guiding architectural and technology choices

Focus 2 → **SOLUTIONS**

SKM: Standardized Knobs, Monitors, Control Algorithms
- Living / designing with unpredictable components
- Large class of runtime solutions for variability and reliability
- Propagating industry acceptance

imec confidential 2007

imec

Knobs & Monitors: how does it work?

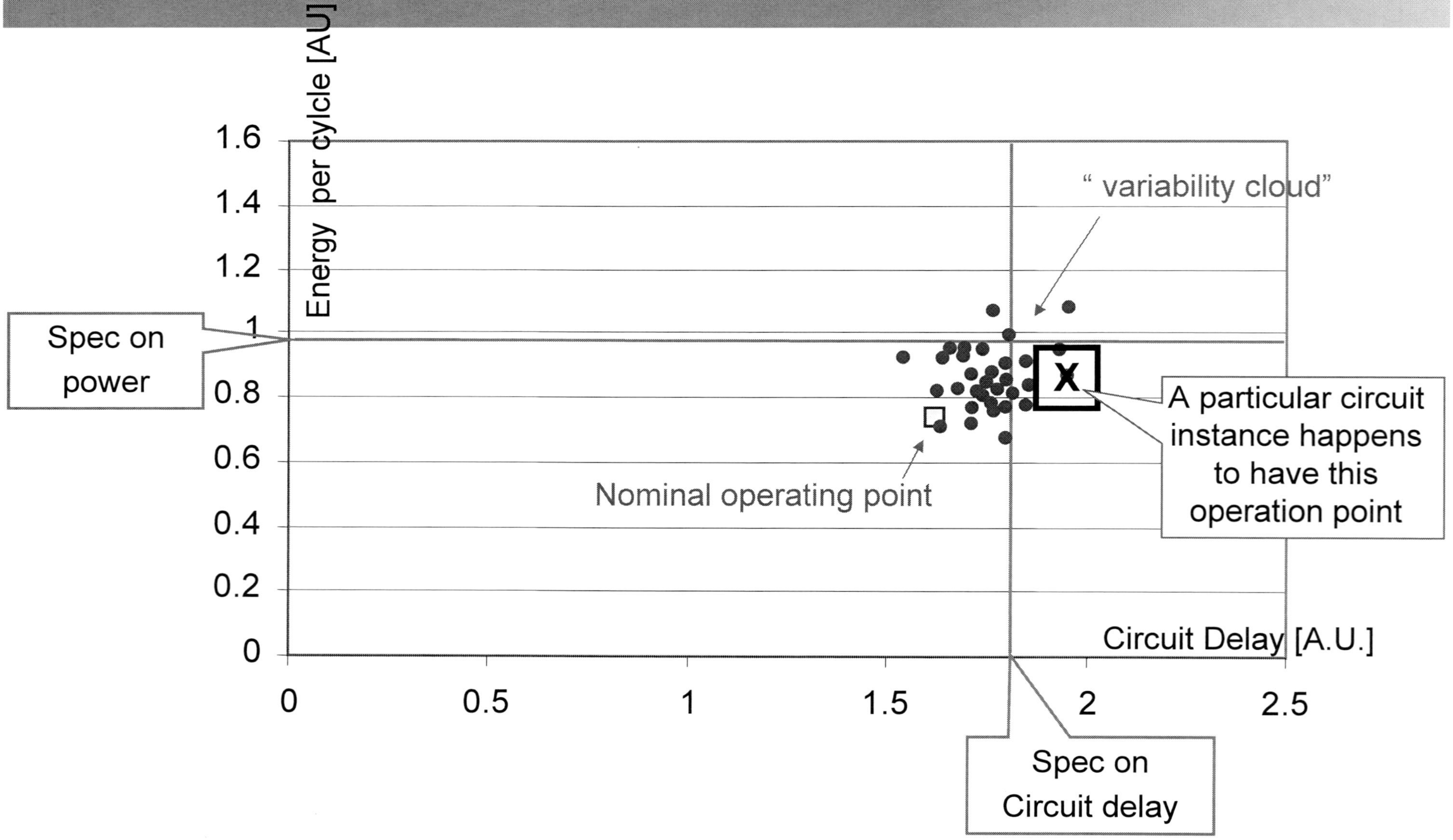

Energy per cylcle [AU]
1.6
1.4
1.2
1
0.8
0.6
0.4
0.2
0
0
0.5
1
1.5
2
2.5
Circuit Delay [A.U.]
" variability cloud"
X
A particular circuit instance happens to have this operation point
Nominal operating point
Spec on power
Spec on Circuit delay

Knobs & Monitors: how does it work?

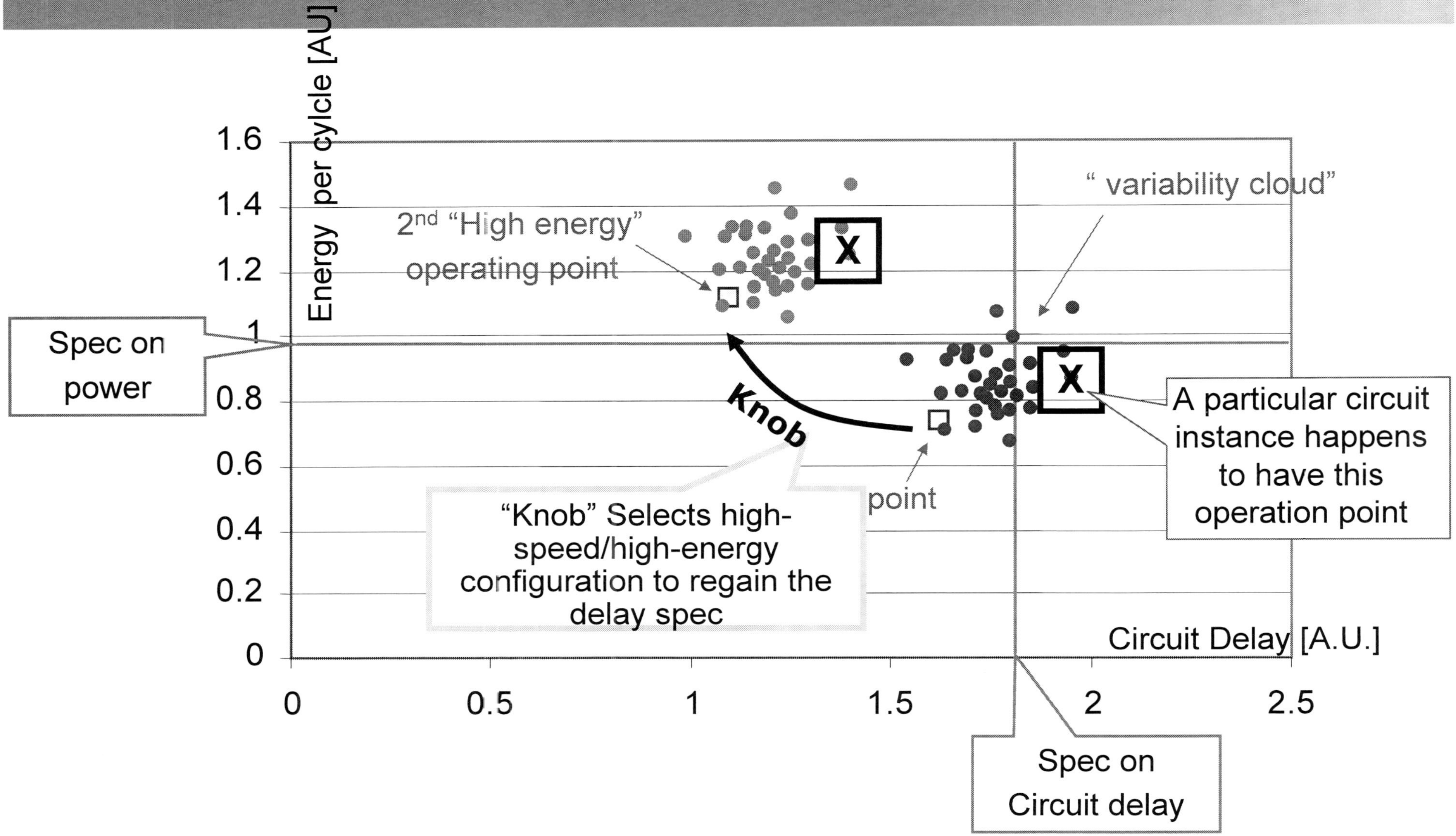

"variability cloud"

A particular circuit instance happens to have this operation point

X

2nd "High energy" operating point

X

point

Knob

"Knob" Selects high-speed/high-energy configuration to regain the delay spec

Spec on Circuit delay

Circuit Delay [A.U.]

Energy per cycle [AU]

Spec on power

imec confidential 2007

Fine-grained Knobs and Monitors in SoC

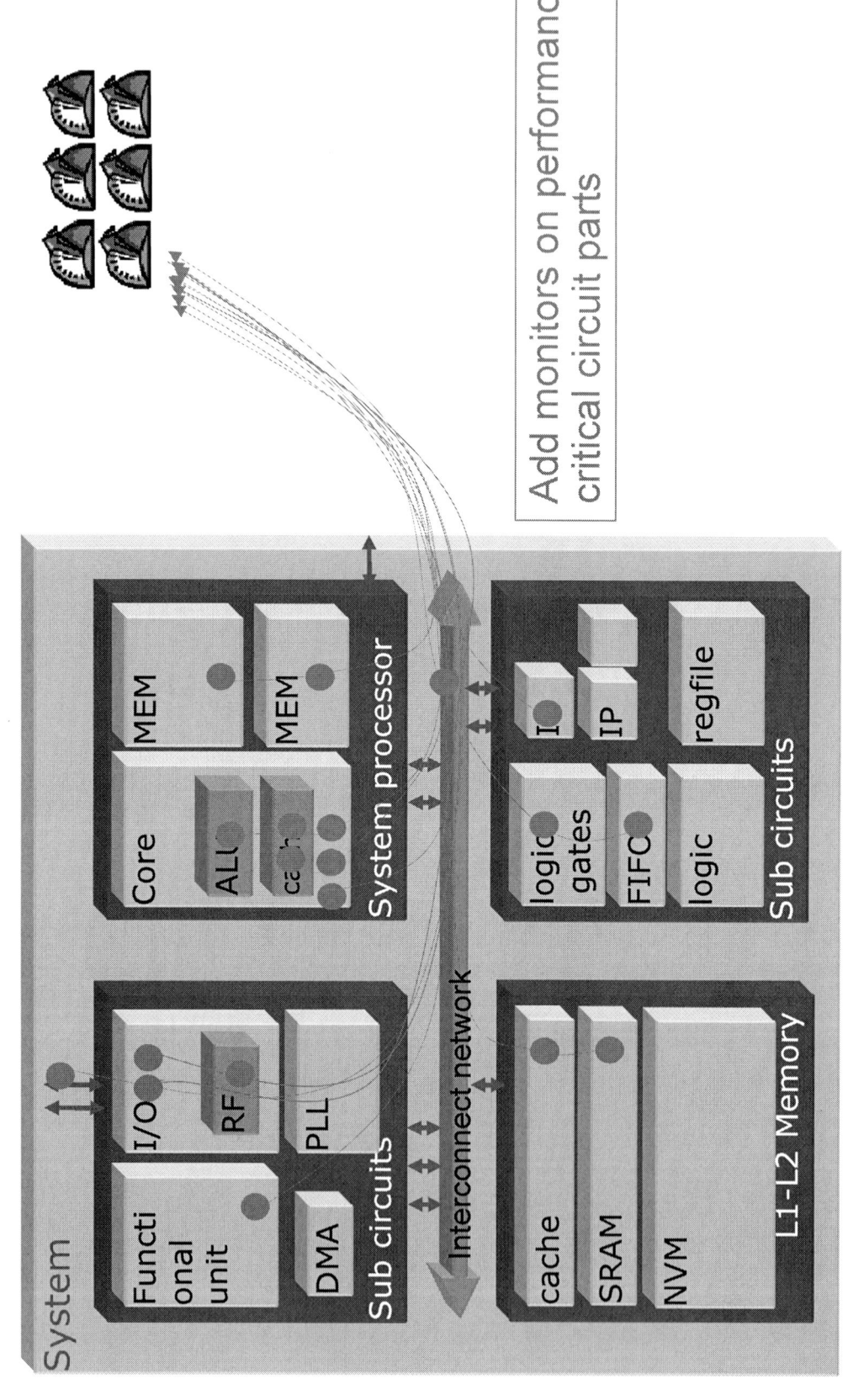

Add monitors on performance critical circuit parts

Fine-grained Knobs and Monitors in SoC

System software

Add monitors on performance critical circuit parts

Add operating point Knobs on performance tunable circuit parts

Put the control intelligence in [embedded] software

System

Core
MEM
MEM
ALU
cache

System processor

I
IP
regfile
logic gates
FIFO
logic
Sub circuits

Functional unit
I/O
RF
PLL
DMA
Sub circuits

Interconnect network

cache
SRAM
NVM
L1-L2 Memory

imec confidential 2007

example: knobs for runtime trade-offs in SRAM

Pre-Charge circuit

SRAM ARRAY

Column Mux

Sense Amplifiers

Output drivers

Output Data

timing ctrl

ROW DECODER

Input Address

Input Address

Column decoder

Timing Ctrl

precharge

Critical driver

Buffers/drivers are present in different parts of memory architecture

Limited impact in area
large impact for energy vs delay

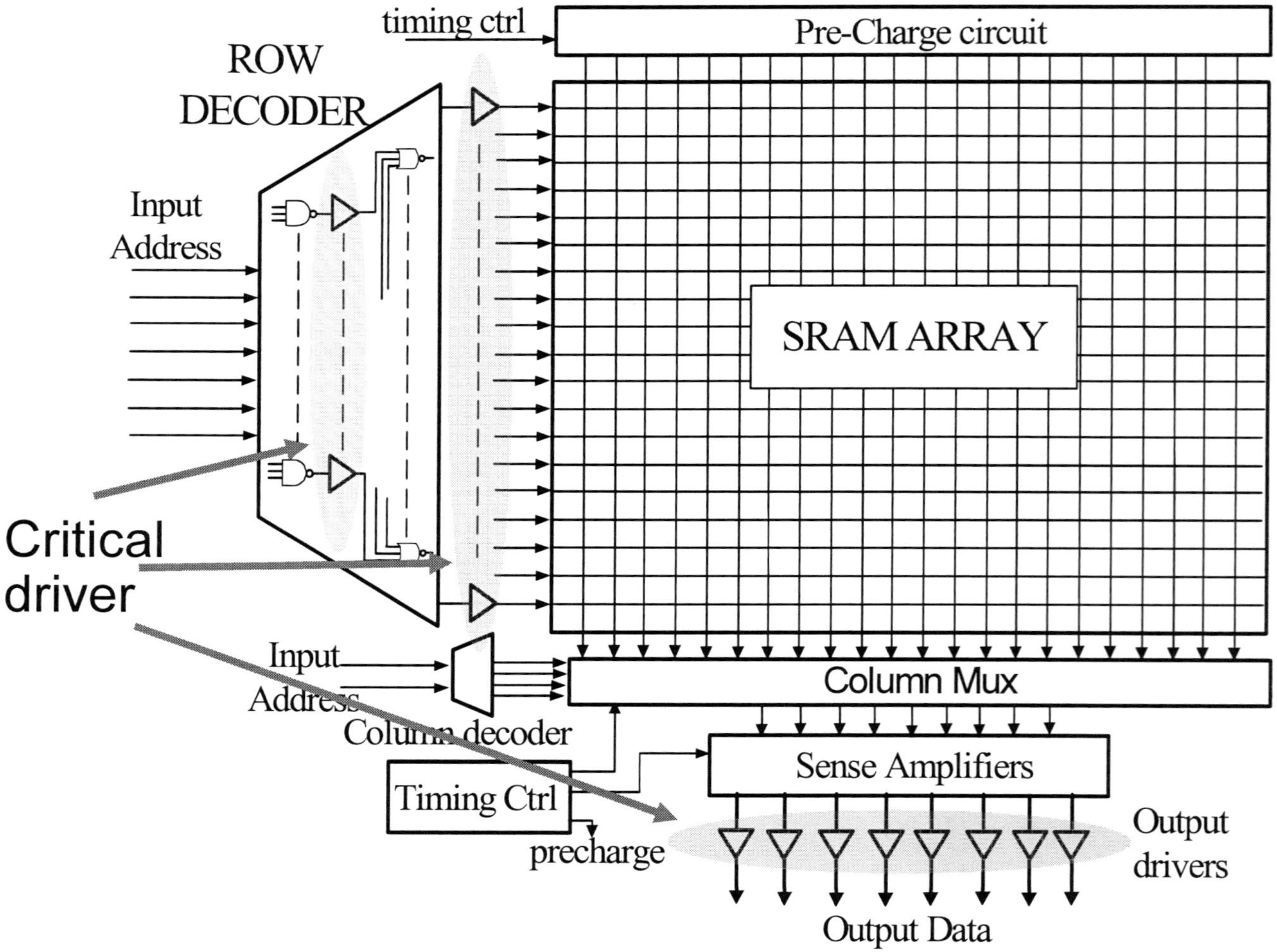

imec confidential 2007

Variability at System level is called... "yield"

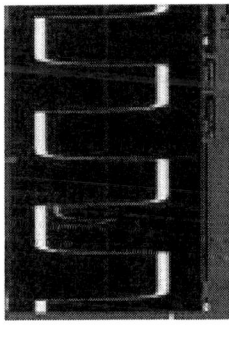

System parametric yield

energy consumption

execution time

Estimation of parametric yield in the energy/timing domain, allowing _what-if_ questions as:

- How much parametric yield for given spec?
- How much worst case and average energy?
- What are the yield critical blocks?
- What-if the architecture changes?

Yield estimator

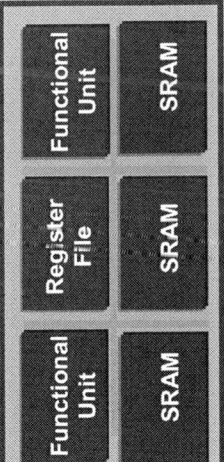

Functional Unit	Register File	Functional Unit
SRAM	SRAM	SRAM

System architecture running application

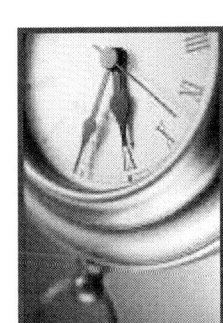

Specifications (power and timing constraints)

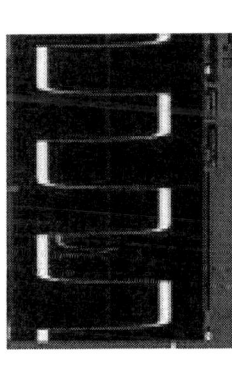

Module-level activity information

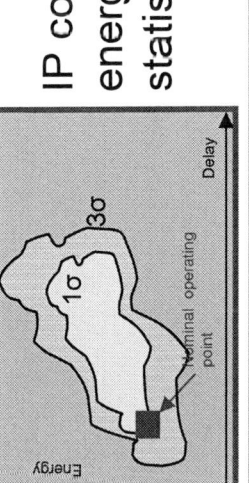

IP component energy/delay statistics

imec confidential 2007

imec

Variability Aware Modeling ... yields "Yield"

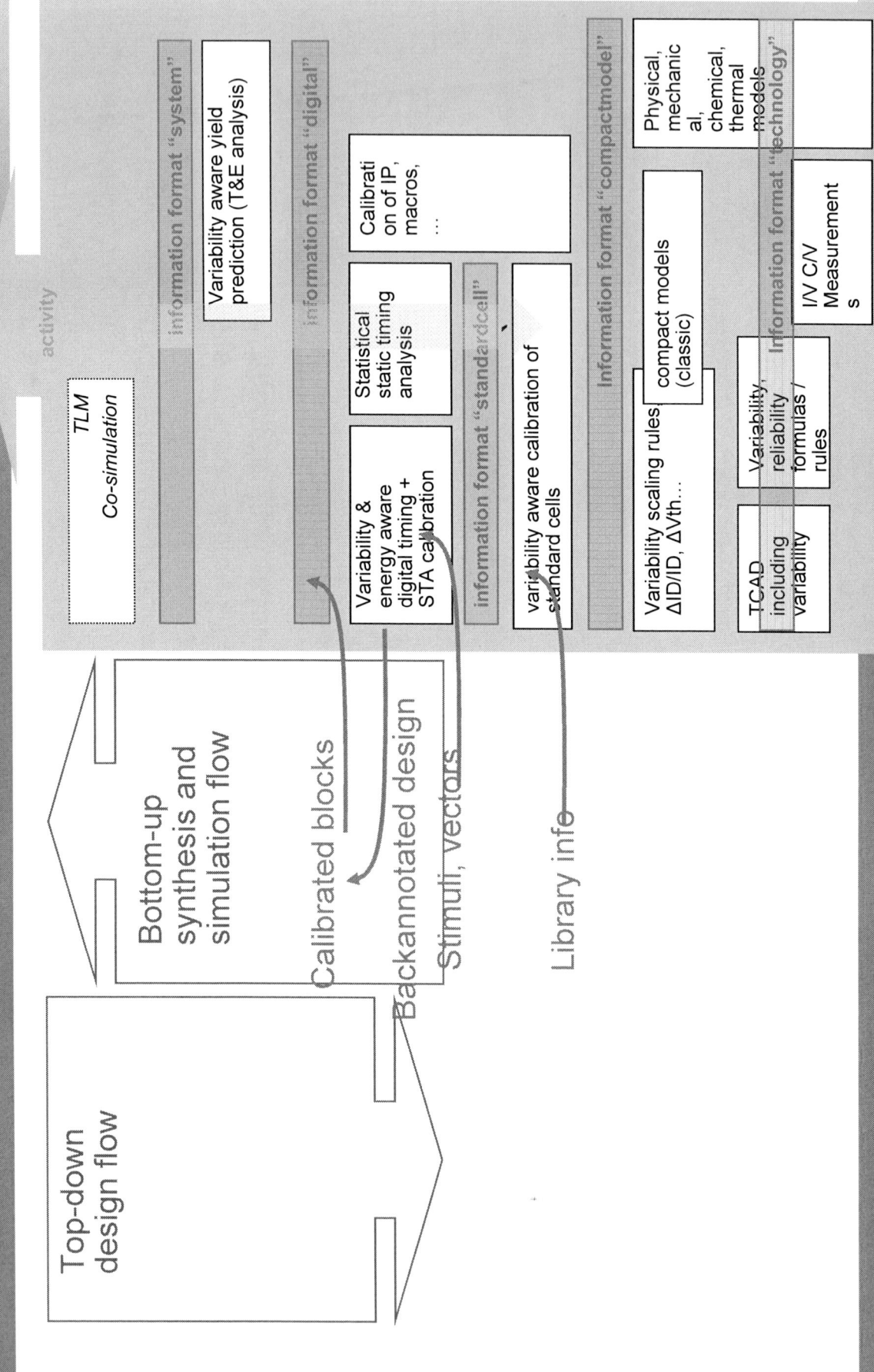

imec confidential 2007

Positioning versus state of the art

TAD – DFM – DFY

- TAD scope is to solve issues at system and circuit level
- DFM/DFY acts primarily at tape-out, Si hardening and mask shop level

TAD – SSTA

- Power and application load are included in TAD not in SSTA
- Propagation of reliability mechanisms

SKM – other circuit methods

- Industry acceptance by standardization; independent developments for knobs (circuits), monitors (circuits) and control (algorithms)
- Generic runtime variability countermeasure method, also able to handle drift, degradation and environmental changes.
- Many specific implementations might coincide with published data

imec confidential 2007

Conclusions
Key benefits of Technology Aware Design (TAD)

What does TAD pursue?

- ## Answer strategic questions
  - tool for trade-off technology and architectural options

- ## Sign off in view of yield:
  - abandon guard band/corner/worst case design
  - go for true yield vs. power vs. performance tradeoff

- ## Offer design level solutions
  - Knobs and Monitors, a Standardized way to realize self-healing circuits
  - overcome the performance and yield loss due to variability, reliability degradation mechanisms

imec confidential 2007

Smartphones and Beyond
Industry Wide Mega-Issues & Mega-Opportunities in the Post-Convergence World

David Wood, EVP Research, Symbian

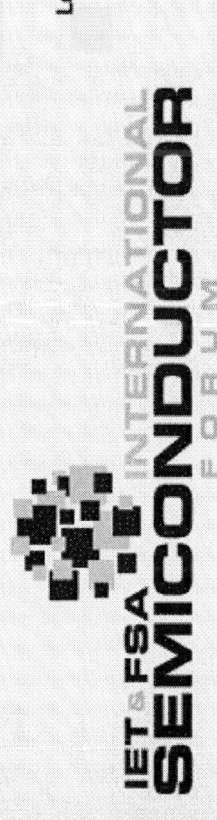

IET & FSA INTERNATIONAL SEMICONDUCTOR FORUM

14 - 15 May 2007
Le Palais Des Congrès De Paris

symbian

Global installed base at end of 2006

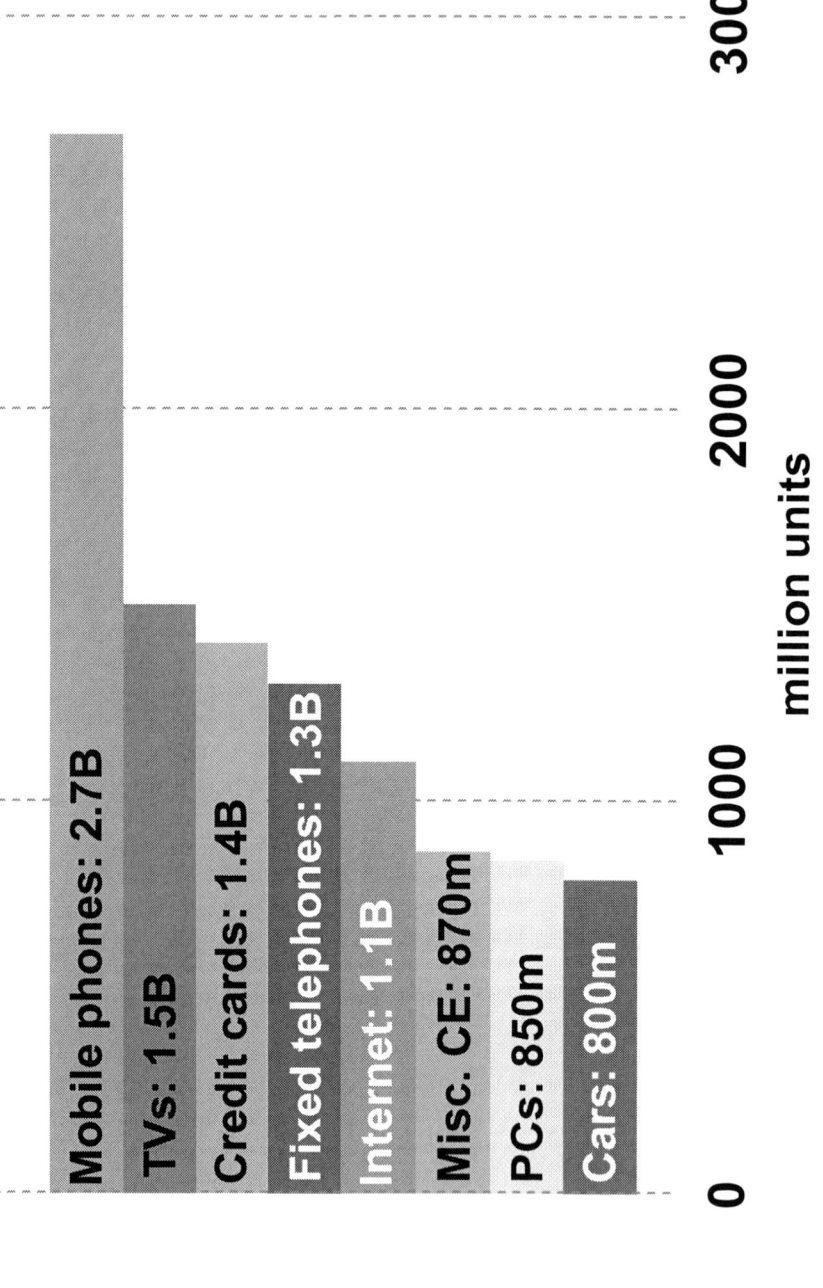

Mobile phones: 2.7B
TVs: 1.5B
Credit cards: 1.4B
Fixed telephones: 1.3B
Internet: 1.1B
Misc. CE: 870m
PCs: 850m
Cars: 800m

0   1000   2000   3000

million units

"Misc. CE" include: PDAs, mp3 players, PVRs, camcorders, game consoles and digital cameras

Source: Communities Dominate Brands – Tomi Ahonen & Alan Moore

symbian

Three waves of mobile phones

Smartphones

The best is still to come

Feature phones

~2005

Voice centric

~2000

Phone functionality

- Rich programmability
- Virtuous cycle
- Innovation
- Applications & services
- *Personal productivity*
- *Business productivity*
- Mobile commerce
- Customisability

- Graphics display
- Colour
- Camera
- Audio: Ringtones+
- Video
- Memory
- Information
- Personalisation

- Great communications
- Voice (& text)
- Pocketability
- Size
- Weight
- Battery life
- Robustness
- Reliability

symbian

The smartphone market open virtuous cycle

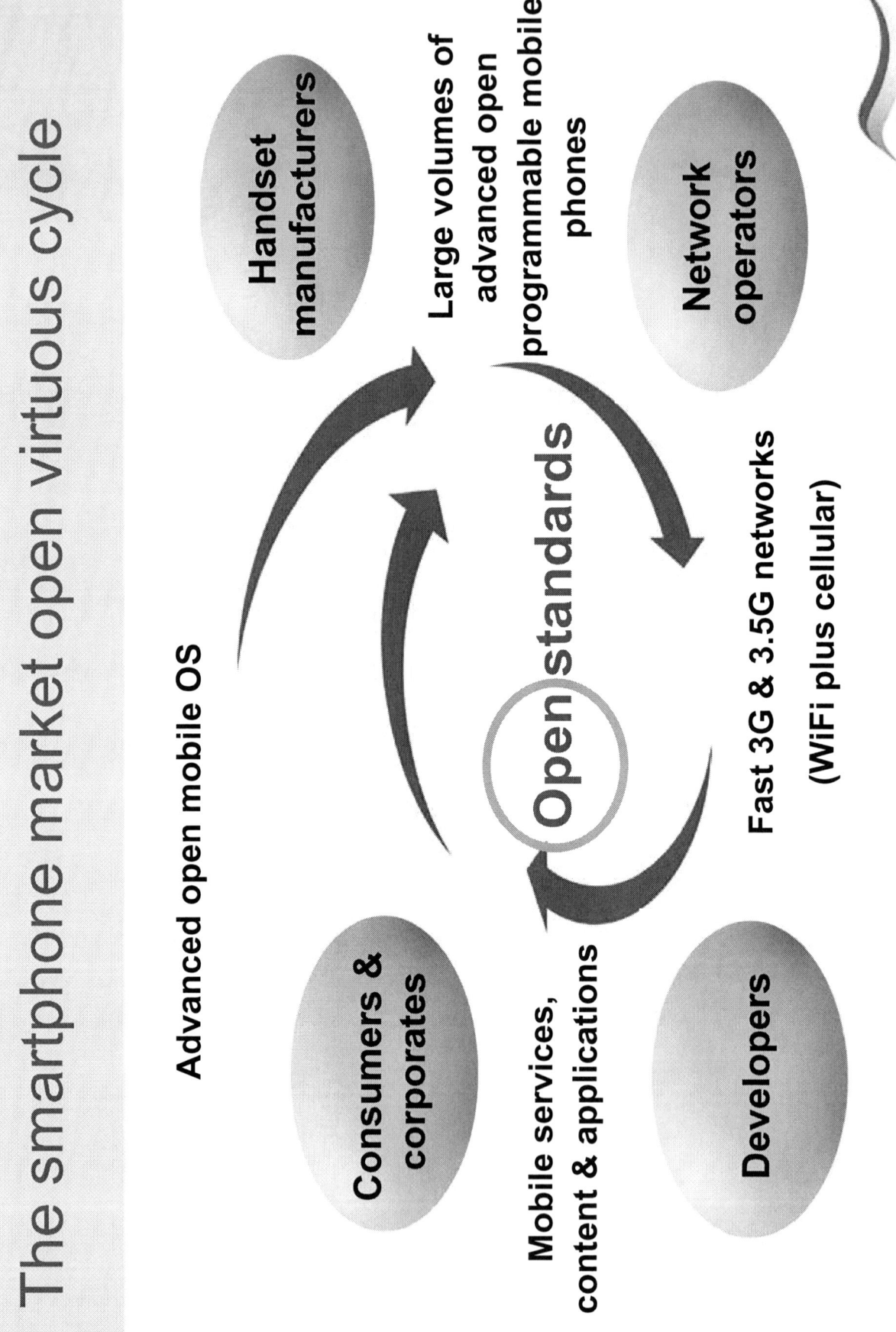

Advanced open mobile OS

Handset manufacturers

Large volumes of advanced open programmable mobile phones

Network operators

Open standards

Fast 3G & 3.5G networks (WiFi plus cellular)

Consumers & corporates

Mobile services, content & applications

Developers

symbian

"Open" means

- Programmable
- Interchangeable
- Collaborative
- Open-minded
- Free-flowing

> Opportunities build on openness

"Closed" means

- Fixed functionality
- Non-standard add-ons
- Overly competitive
- "Divine right" attitude
- Bottlenecks and chokes

> Risks threaten a resurgence of closure

symbian

The smartphone market open virtuous cycle

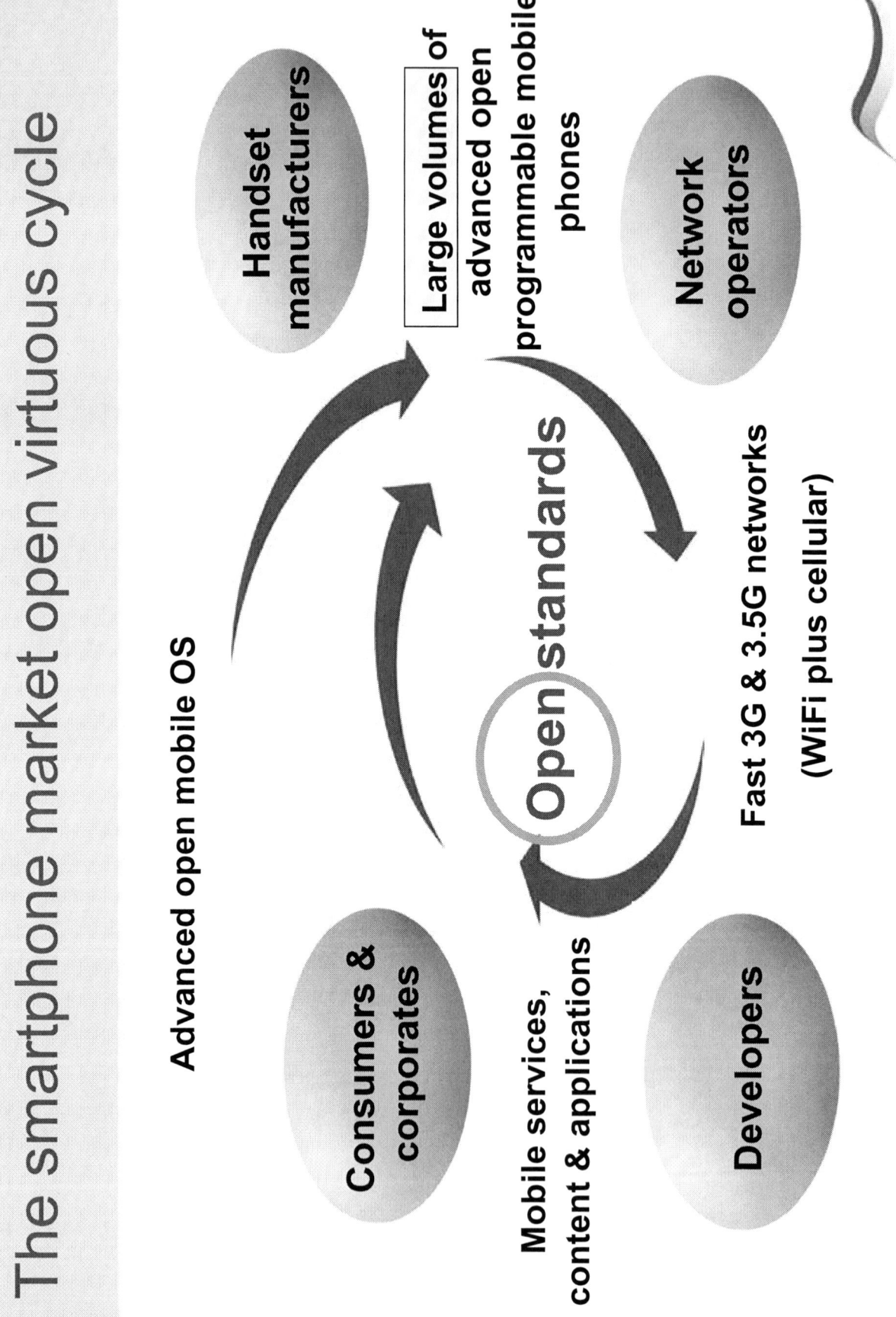

Handset manufacturers

Large volumes of advanced open programmable mobile phones

Network operators

Advanced open mobile OS

Open standards

Fast 3G & 3.5G networks (WiFi plus cellular)

Consumers & corporates

Mobile services, content & applications

Developers

symbian

Symbian smartphone shipments

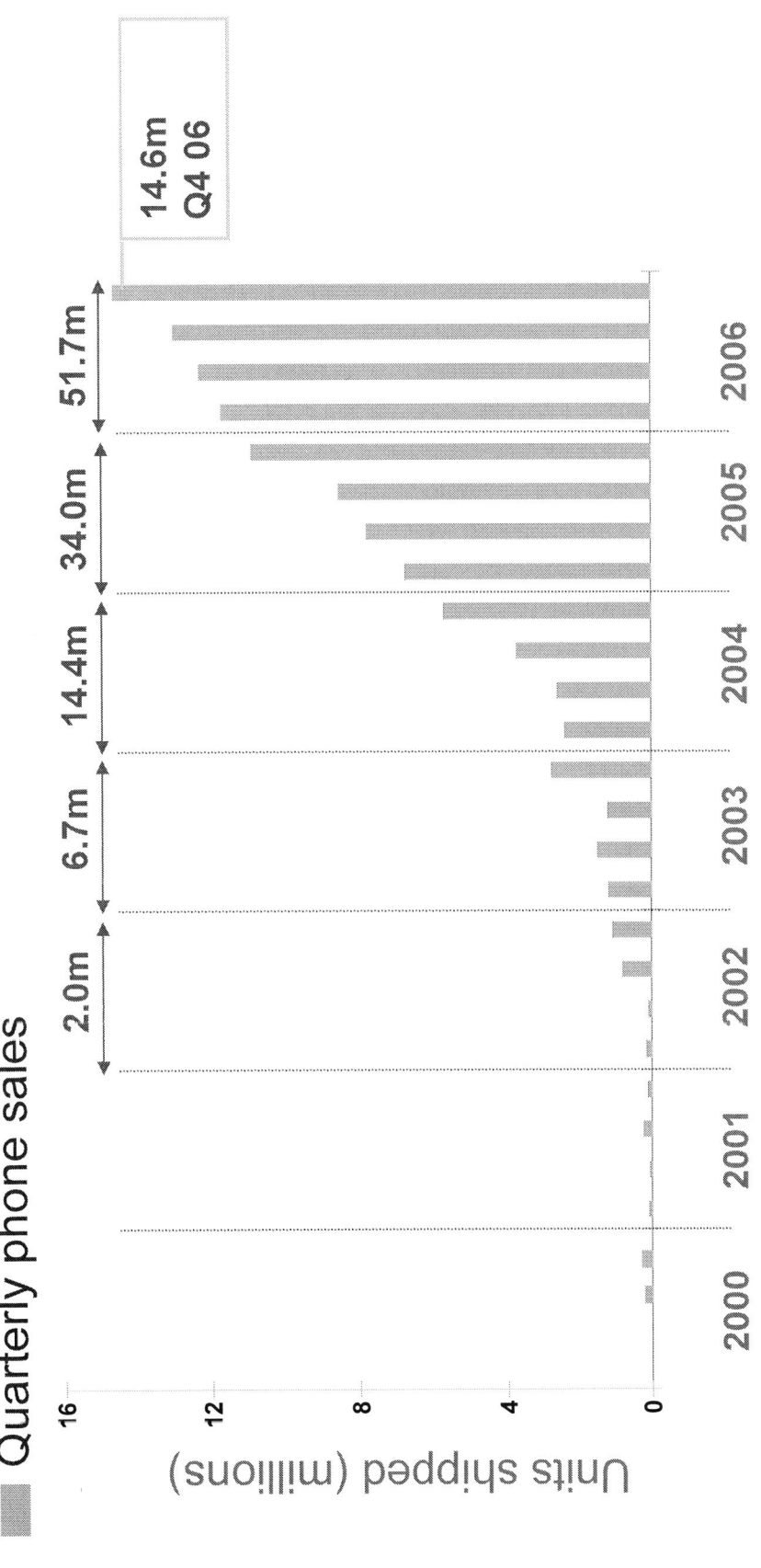

110 million cumulative units shipped at the end of Q406

symbian

Sustained smartphone market share

Source: Canalys

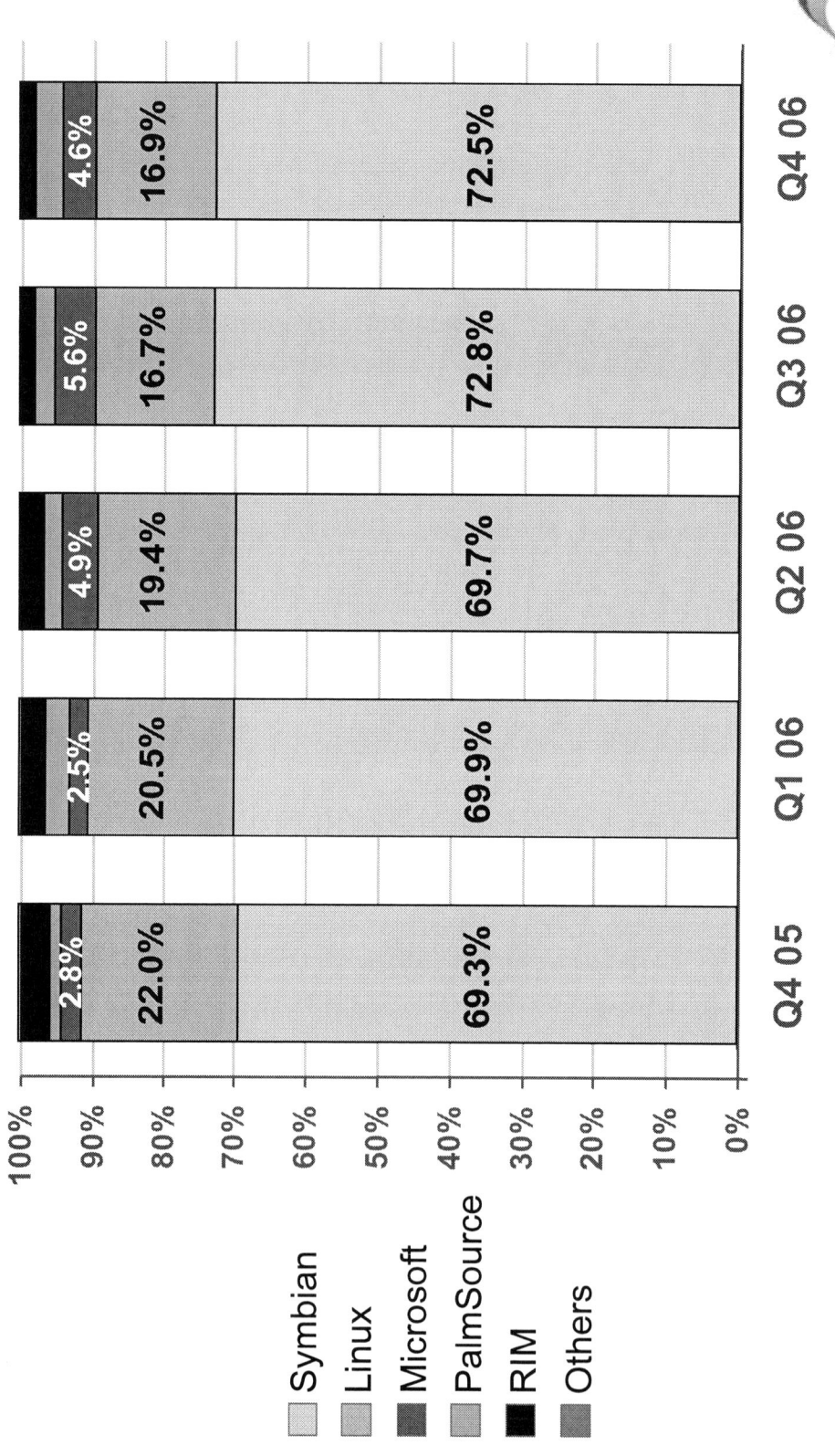

symbian

Regional smartphone sales breakdown (Q4 06)

Source: Canalys

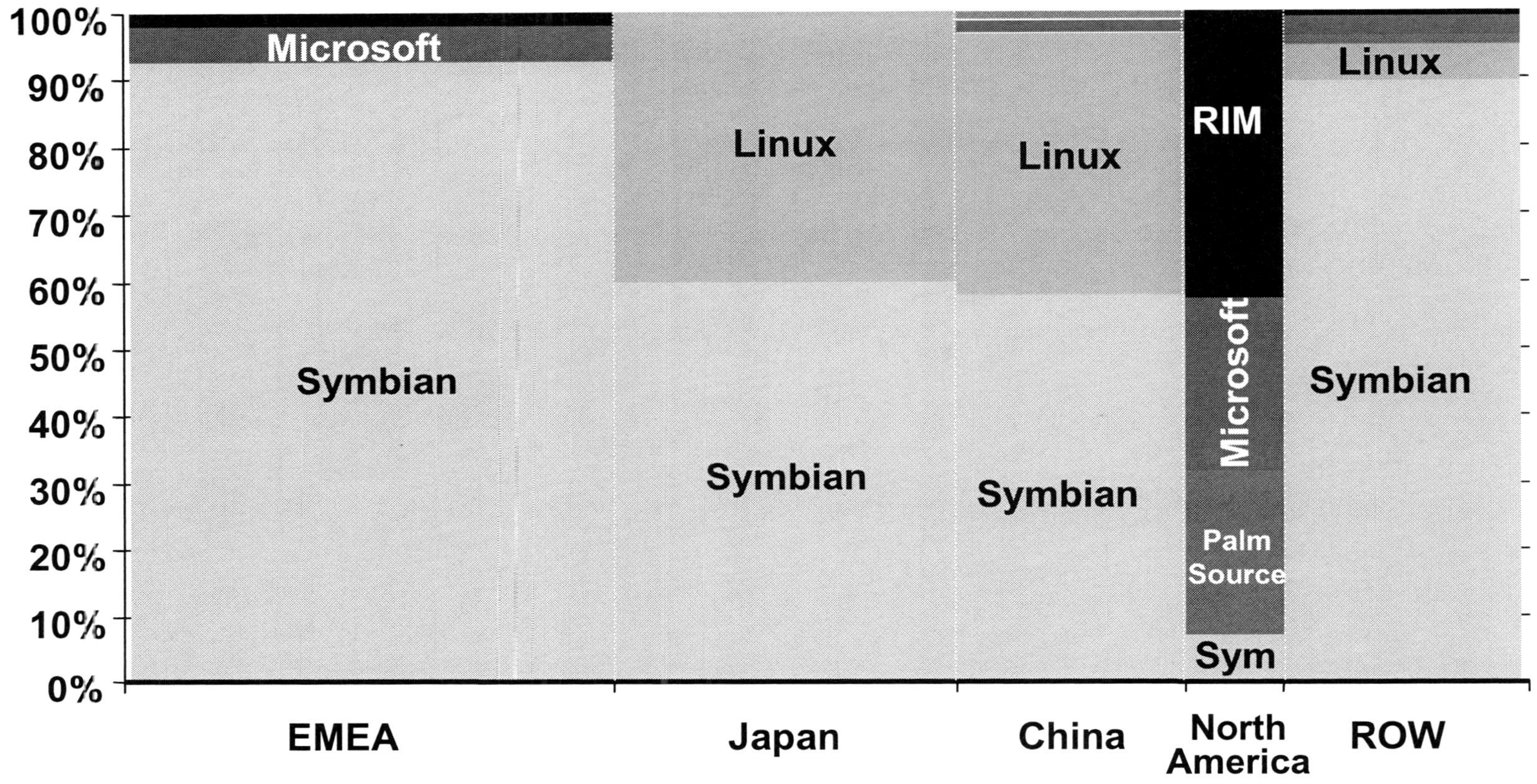

Total Q4 2006 smartphone sales: 20,217,920

symbian

Smartphone Bill Of Materials cost decline

Notional threshold BOM

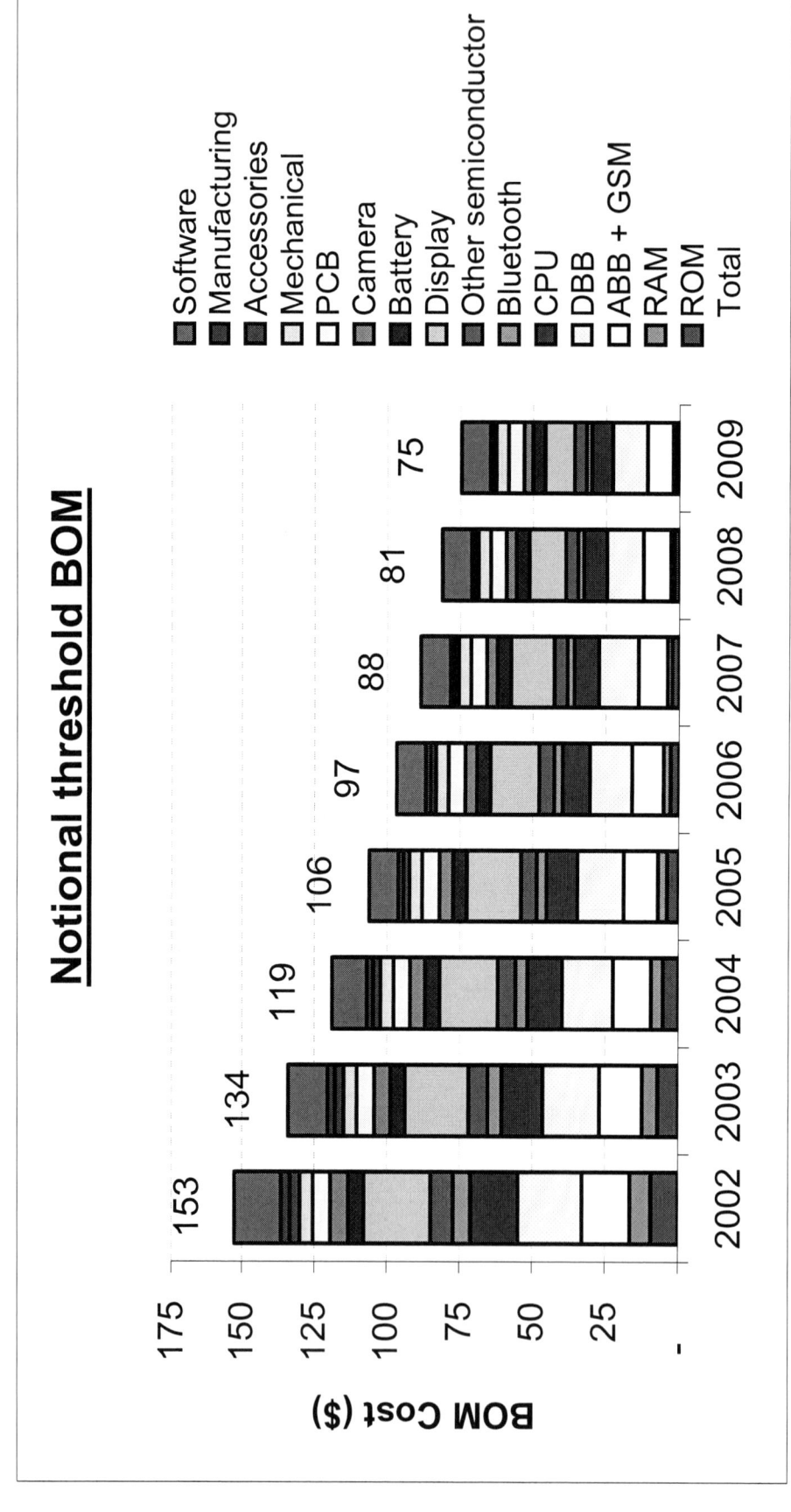

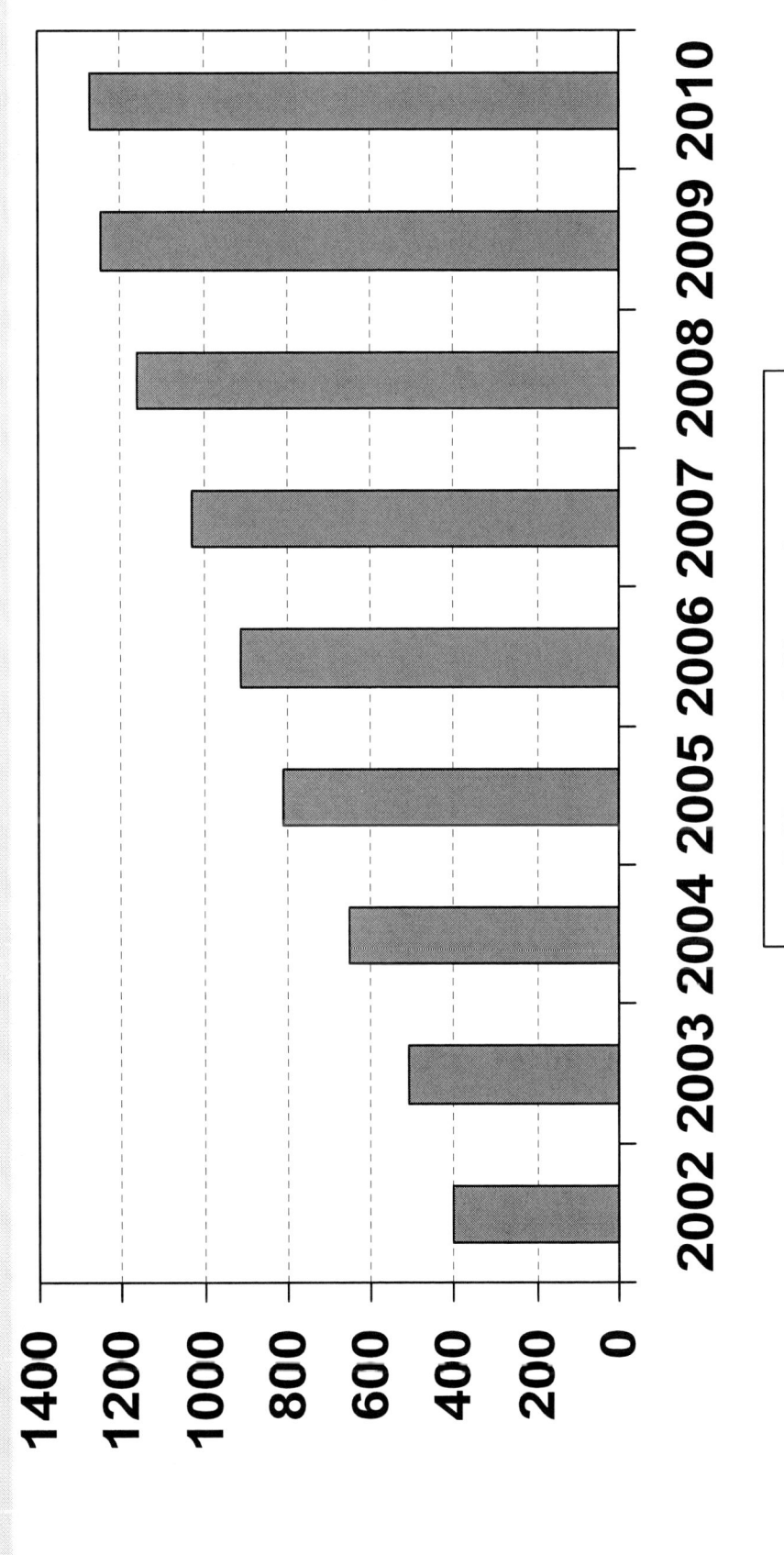

Smartphone _addressable_ market per annum (M)

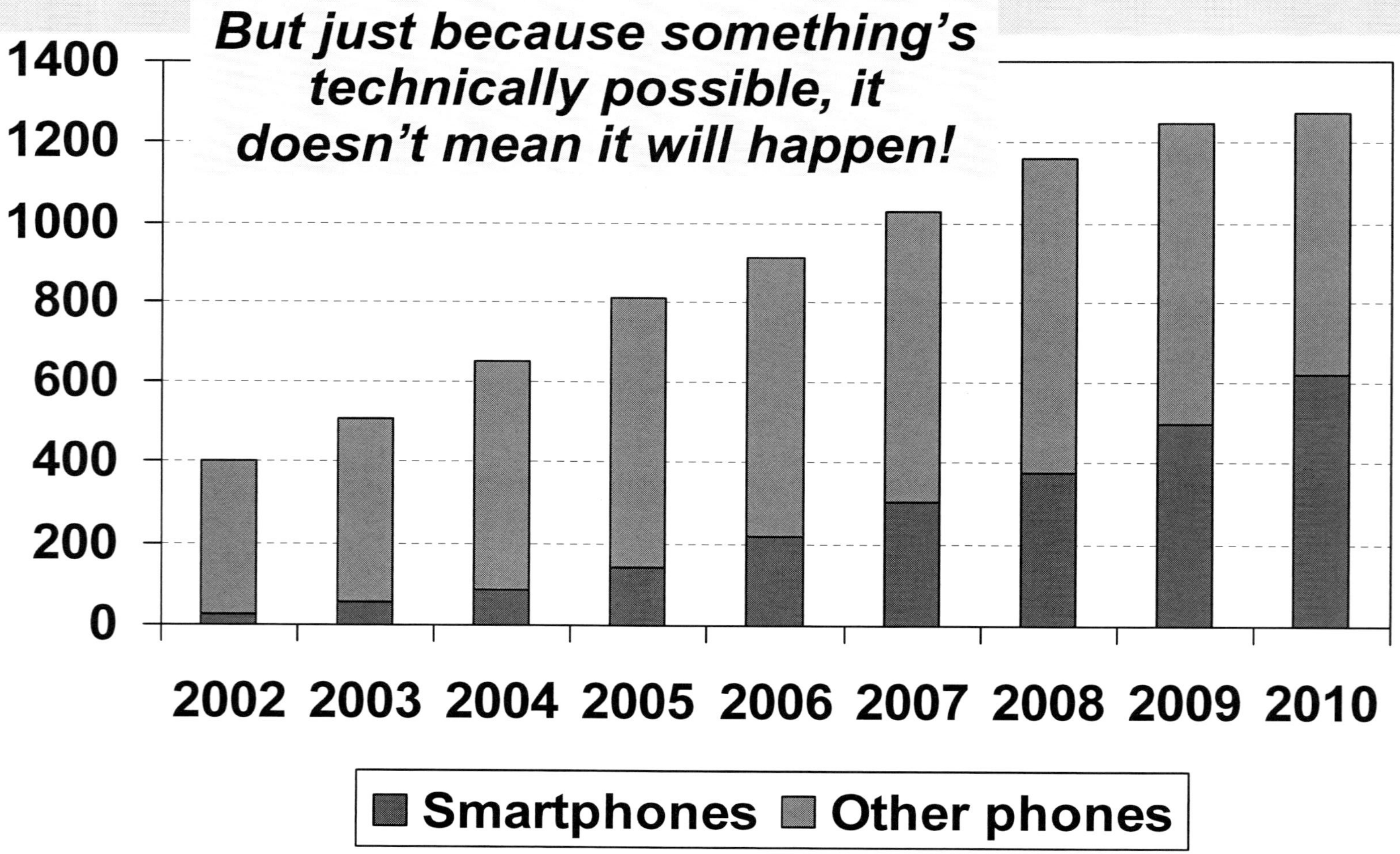

But just because something's technically possible, it doesn't mean it will happen!

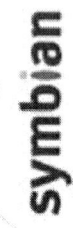

symbian

Technology adoption life-cycle

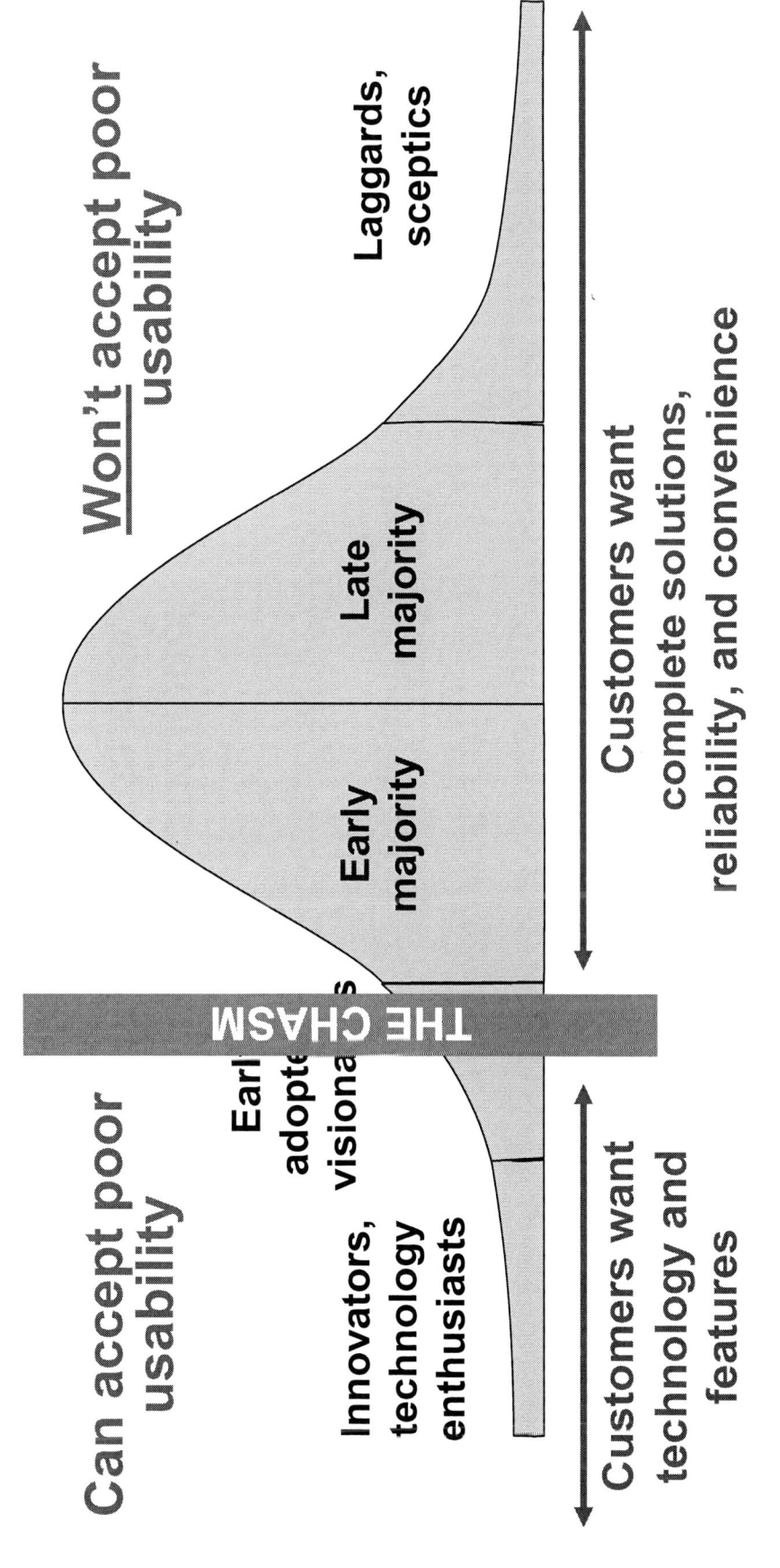

Can accept poor usability

Won't accept poor usability

Innovators, technology enthusiasts

Early adopters, visionaries

Early majority

Late majority

Laggards, sceptics

THE CHASM

Customers want technology and features

Customers want complete solutions, reliability, and convenience

(source: Geoffrey Moore)

symbian

Smartphones will cross the chasm if:

- They allow users to build on & do more of the things that caused users to buy phones in the first place

  - …Communication (and messaging)

  - …Safety & connection (timely info in context)

  - …Fashion & fun (personalisation)

- AND they allow users to do these things simply & swiftly

  - …(Even though the phones themselves are increasingly complex)

- AND on that basis, provide additional functionality of genuine value to users

  - …For example, becoming people's preferred personal mobile gateway into the digital universe…

symbian

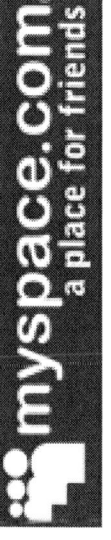

Mobile gateway into the digital universe

All company and product names & logos are registered trademarks of their respective holders

Accessing the digital world, while mobile

- Improved bandwidth – faster wireless networks

- Improved display technology

  - More pixels, higher resolution, more colours, sharper screens

- Enhanced UIs, keyboards, HWR, auto-complete

  - Easier for users to enter data

- Predictable, flat-rate data charging systems

  - Removing users' fear of unexpectedly large phone bills

- Intelligent web browser software!

  - Page re-layout, intelligent proxies, incremental redrawing...

- Intelligent non-browser software!

  - Dedicated native applications...

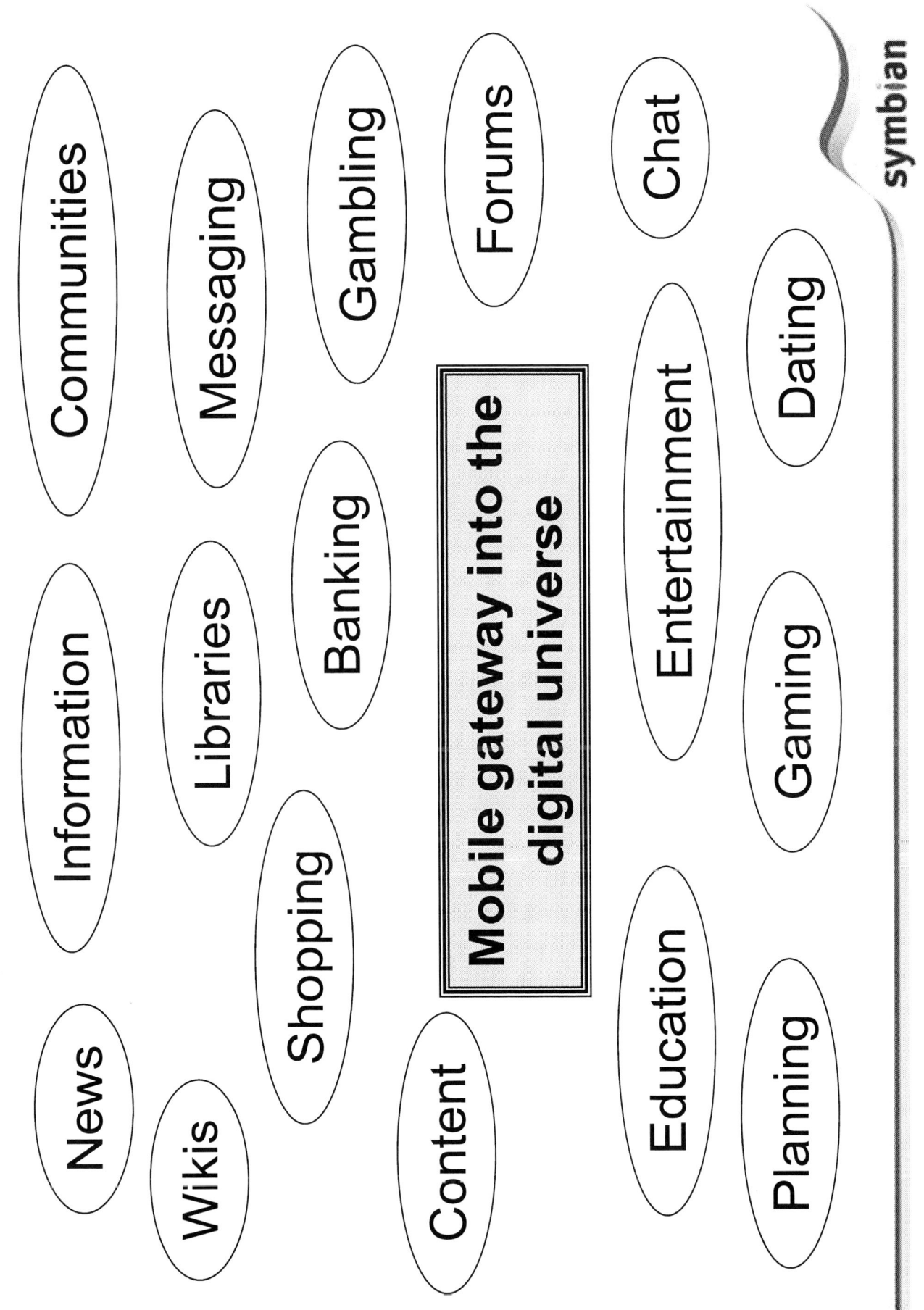

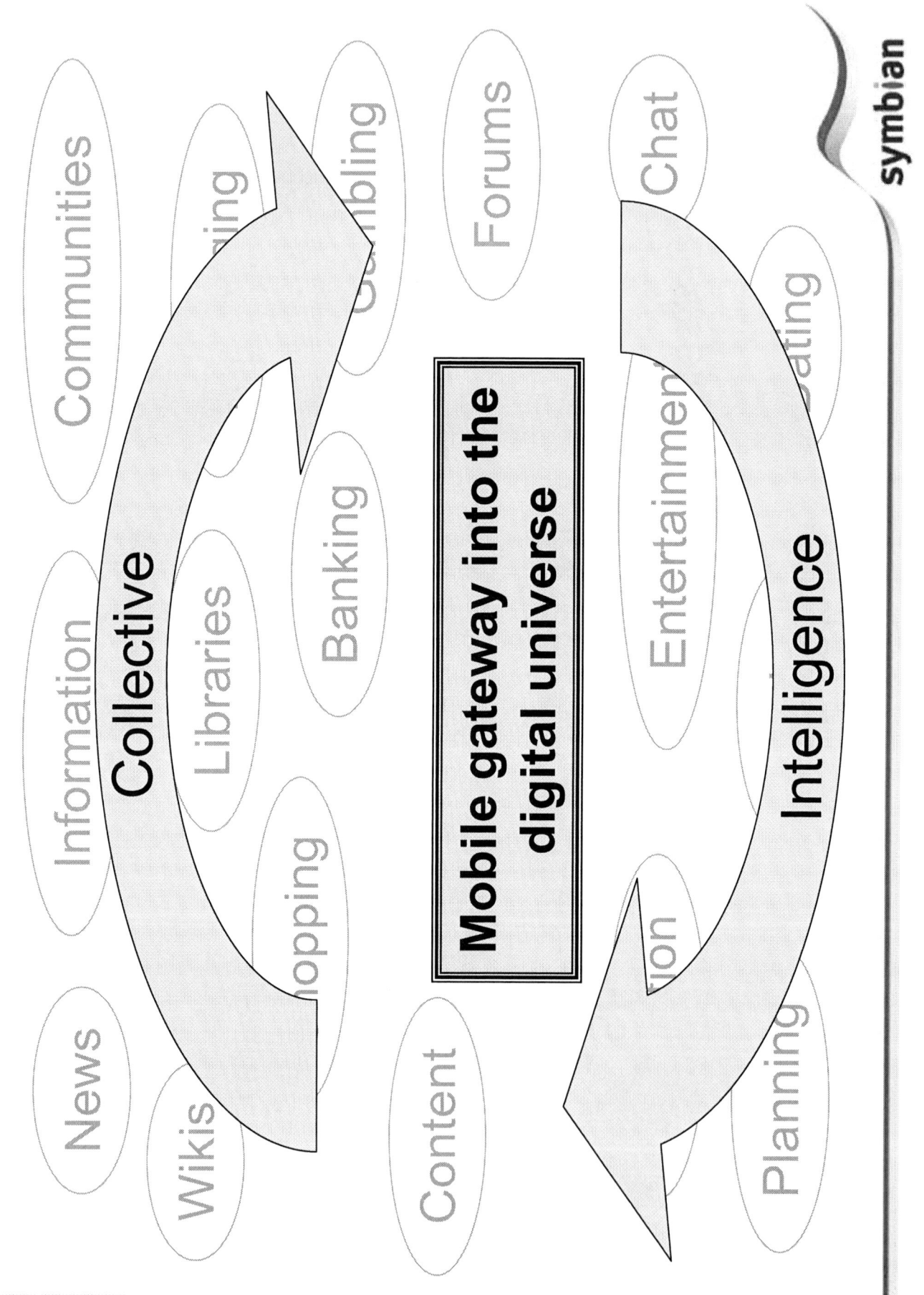

Convergence: *Novelty via unexpected convergence*
Smartphones as pocket melting pots

Newspapers

Tickets

Camcorder

Dictionary

Music player

Games console

Alarm clock

Camera

Radio

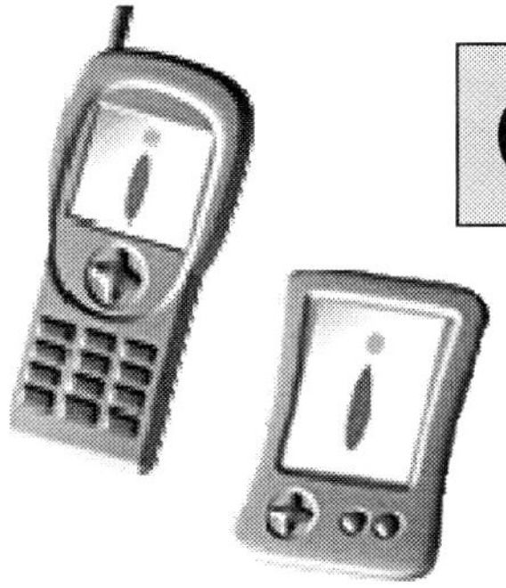

BlackBerry

Web tablet

Health monitor

Diary

Keys

PDA

Books

Calculator

To-do list

Vouchers

Wallet

TV

Map

Watch

Smartphones everywhere

symbian

Vision: Analysts' forecasts

Andrew Brown, mobile devices specialist at <u>IDC</u>

- **Cumulative sales of smartphones will reach over 1 billion units by 2011**

- The smartphone segment of the handset market is seeing strong annual growth and is expected to rise from 57 million units in 2005 to around 250 million units by the end of 2010

- The advent of single chip designs will attract a growing number of licensees and drive greater penetration into the mid-market

Stuart Robinson, Director, <u>Strategy Analytics</u>

- **Cumulative sales of smartphones will reach 1 billion units by the first quarter of 2011**

- The smartphone segment of the handset market is expected to rise from 45 million units in 2005 to around 300 million units by the end of 2010

Nick Spencer, Analyst, <u>Canalys</u>

- **Global shipments of smartphones will reach 1 billion by 2012**

- We are still at an early stage of market development, with businesses & consumers only scratching the surface of what is and will be possible with such devices

Driving costs down and volumes up

- Need to drive down costs:
  - ... Device build costs, and service deployment costs, and usage costs

- Costs are reduced, in part, by the working of Moore's Law

- Moore's Law is driven by ingenuity in real companies
  - ... Eg ARM **ARM** THE ARCHITECTURE FOR THE DIGITAL WORLD®

- An even bigger effect on costs (and hence on volumes) is
  - ... Economies of Scale, and *the Learning Effect*, and the Free Market

- "Practice makes perfect"
  - ... Companies will practice very hard, if the potential rewards are very high
  - ... The most important driver for the increased value of an ecosystem platform is the volume sales of the ecosystem

symbian

Vision:
Evolution of target device for Symbian OS

1986   **1996**   **2006**   **2012…**

PDA

- Electronic organiser
- Interactive manipulation of local data
- Battery life
- Memory constrained
- Instant-on
- Fast task-switching
- Graphical screen (overlapping windows)
- RAM persisted application state
- PC Connectivity

Smartphone

- Phone as the most important application
- Smartphone as "Phone Plus…"
- Smartphone as "simply great phone"
- Device start-up
- Cellular baseband abstraction
- Flash persisted state
- IP connectivity and networking

Wirelessly networked mobile multimedia device

- Post telecoms convergence
- All-IP wireless broadband and multiple bearer management
- Real-time networking
- Multimedia (creation, consumption, communication)
- Graphics and multimedia hardware acceleration
- Large fast persistent storage
- Content protection/rights management
- Energy & thermal constraints

symbian

Six horsemen of the apocalypse

- Six seismic challenges standing the way of smartphones fulfilling their promise of another ten-fold capability growth

- Challenges which the entire industry needs to address and solve wisely and collaboratively

symbian

#1: Fire

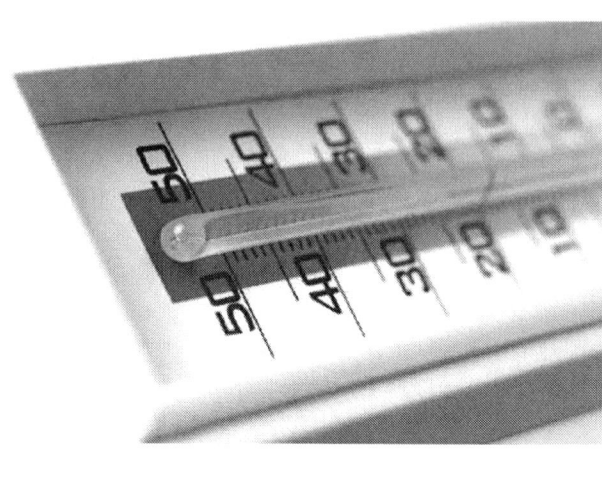

- 2006: Laptop batteries catching fire
- Excess power consumption gives off too much heat
  - ... Laptop clock rates can't keep increasing
- Smartphones have no air-conditioning fans (rightly!)
- **Huge importance of power-handling and power-optimisation**
- Enormous value of an OS that:
  - ::: Focuses on power efficiency
  - ::: Heritage of "small"
  - ::: Supports distributed (parallel) processing
  - ::: Features highly real time characteristics

symbian

#2: Floods

- We need to cope with potential floods of software
  - Software bloat
- Nathan's Law
  - "Software is a gas" – "It expands to fill its container"
- Increasing software bulk kills development projects
  - Windows Longhorn
- You can't cope with floods of software just by applying floods of people (or floods of eyeballs)
  - Too many cooks spoil the broth ("mythical man-month")
- Brooks' Law
  - "Adding manpower to a late software project makes it later"

symbian

Ruthlessly simplify

> Especially when constrained

- **Software inevitably tends towards greater complexity**
  - This is the second law of thermodynamics for software
  - (The first law is that software inevitably tends to grow in bulk)

- **Complexity has n-squared effects**
  - Many more relations to understand

- **Keep seeking to simplify the design**
  - Well-defined classes
  - Expose fewer APIs rather than many APIs
  - Avoid long inheritance trees; Avoid using inheritance for "cleverness"
  - Re-factor designs on a regular basis, wherever possible

- **Keep seeking to simplify the development process**
  - Re-factor the development process on a regular basis

- **Note – make things as simple as possible, but no simpler**

Handling complexity

- Without taming complexity:
  - Projects suffer
  - Users suffer!

- You can't solve it just by applying people and/or money

- You must get the **core system architecture right**
  - You must have the right approach to crafting software

- Increased focus on
  - Tools and documentation
  - Reference designs
  - Skilled consultancies
  - Interface management – binary compatibility
  - Excellent APIs – clear boundaries

symbian

Phone manufacturer adoption life-cycle

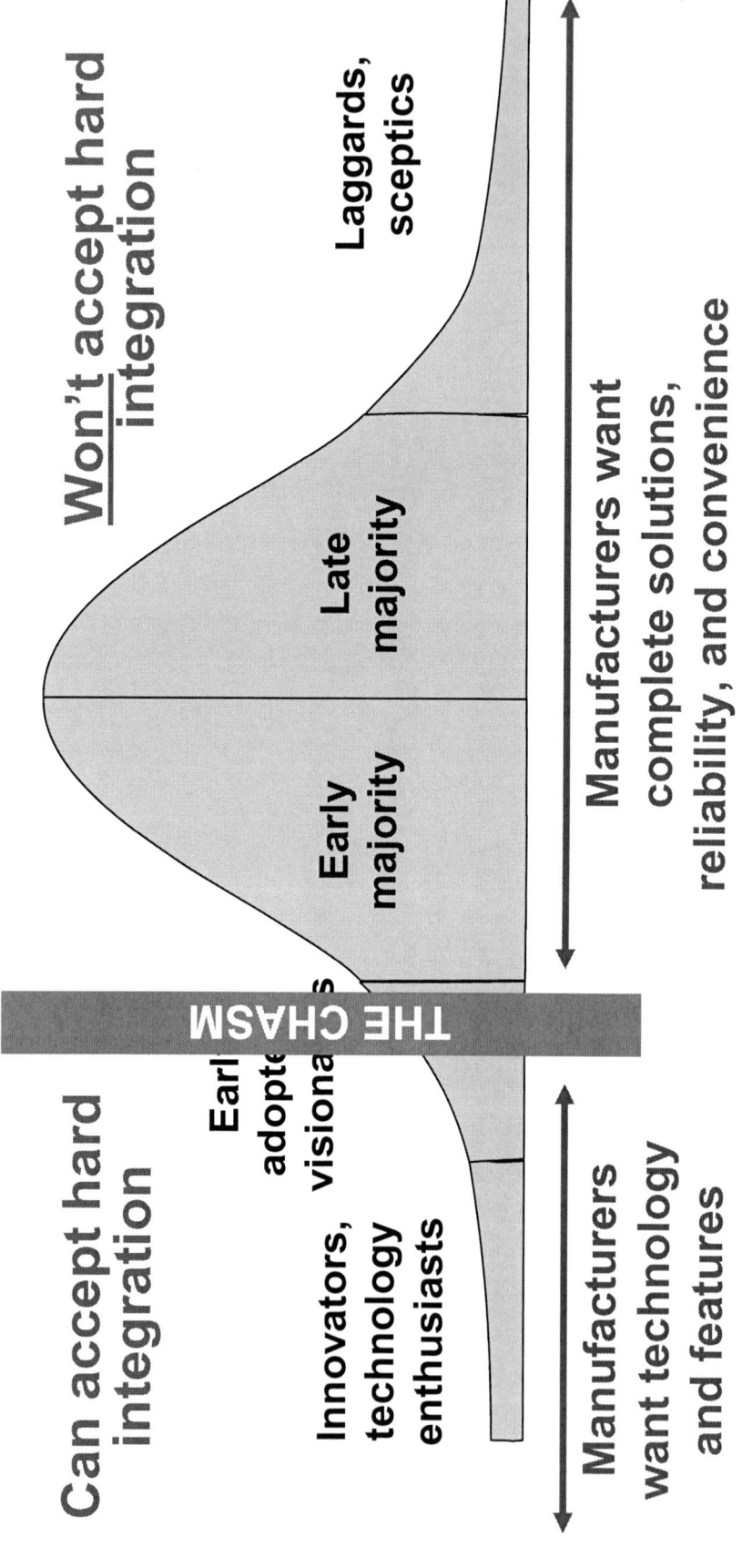

Can accept hard integration

Won't accept hard integration

Innovators, technology enthusiasts

Earl adopte visiona

Early majority

Late majority

Laggards, sceptics

THE CHASM

Manufacturers want technology and features

Manufacturers want complete solutions, reliability, and convenience

(source: Geoffrey Moore)

symbian

#3: Plagues

- Beware plagues of viruses and spam

- Poor security will lead to the disintegration of the network

- *Fear of* poor security will lead to network owners locking down their networks

  - Preventing the open introduction of innovative new solutions

- You have to put security at the heart of your architecture

- Symbian OS v9.x **platform security**

  - Goal: the security works even without the user having to understand it

  - Addresses malware that works using "social engineering"

  - Lightweight but effective: **unique secure rich openness**

The smartphone market open virtuous cycle

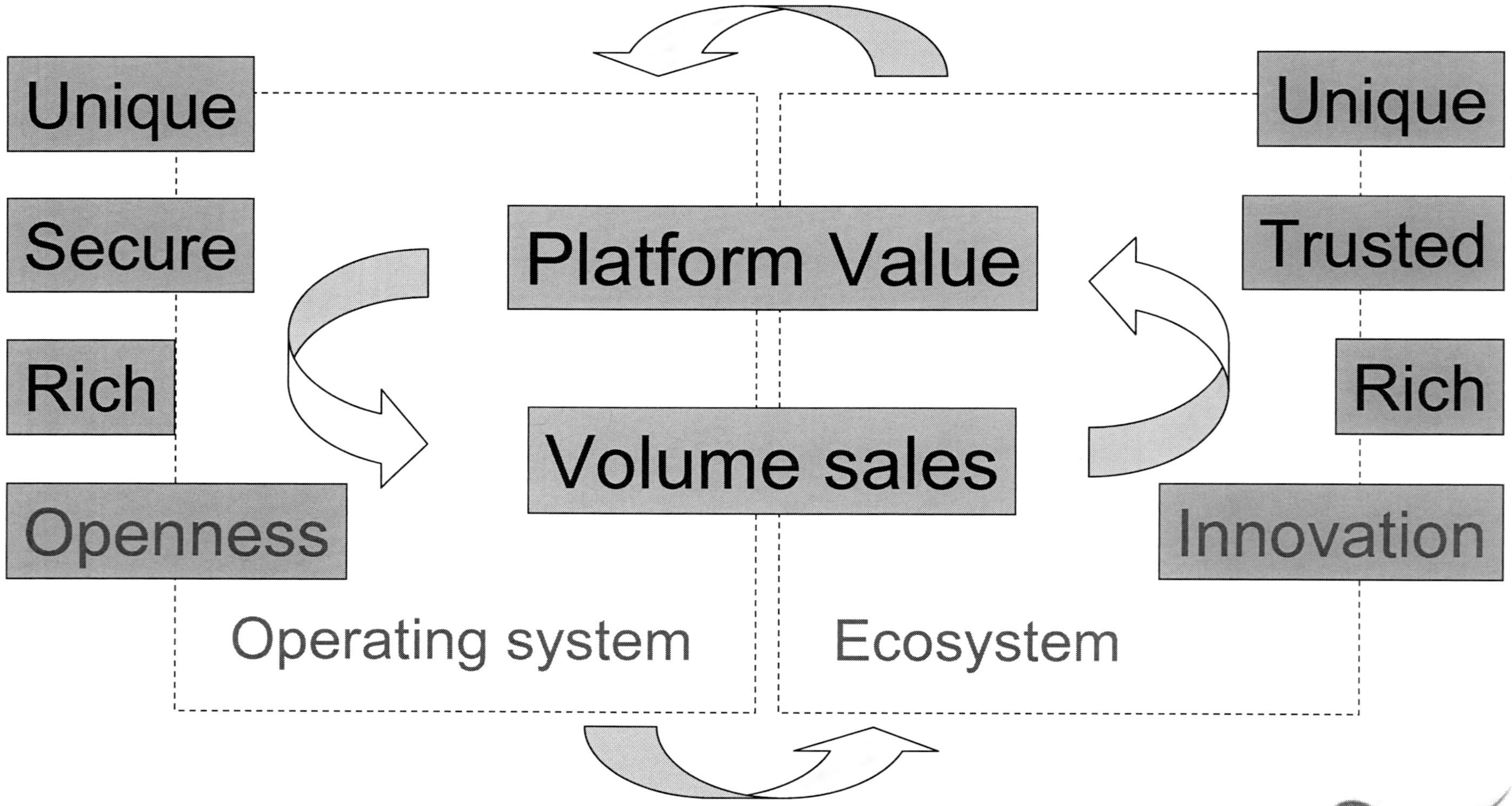

Unique
Trusted
Rich
Innovation

Unique
Secure
Rich
Openness

Platform Value

Volume sales

Ecosystem

Operating system

symbian

#4: Warfare

- Too much smartphone competition will destroy the new virtuous cycle before it can reach its greatest effectiveness

- We want to move from the proprietary, closed world
  - ... With too many solutions being "point-optimised solutions"

- To the open, standards-based, collaborative world
  - ... Where novel solutions are inter-changeable

- We need platforms with rich openness
  - ... Rich openness enables rich innovation in the ecosystem

- **Collaborate before competing**
  - ... Make the pie larger, before we squabble over dividing it up

"Symbian": basis for collaborative partnership

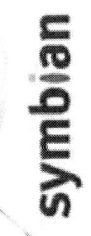

Demi-Moore's Law (network effects)

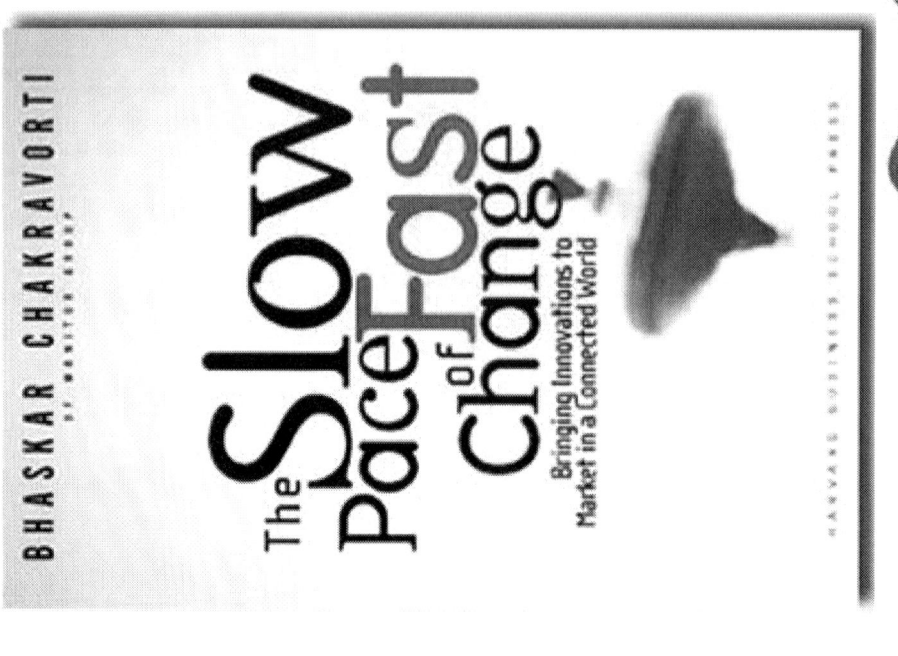

- Disruptive change takes twice as long as Moore's Law predicts?!
  - ...To start with!

- "The slow pace of fast change
  - ...Bringing innovations to market in a connected world"
  - ...By Bhaskar Chakravorti

- Disruptive innovations have to precipitate the dismantling of an existing equilibrium
  - ...And help orchestrate the transition to a new equilibrium
  - ...Supply the "activation energy"

#5: Greed

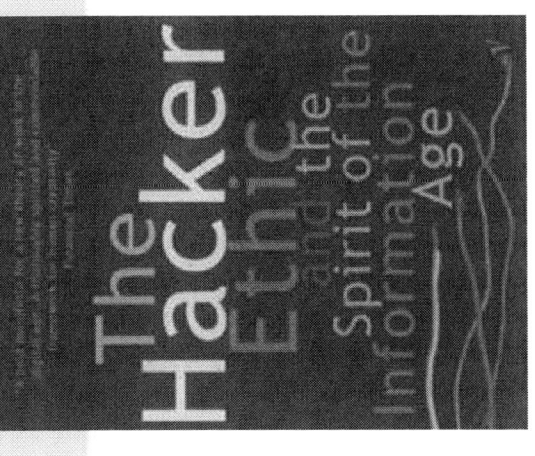

- Companies are motivated by commerce
  - "The business of business is business"
  - Money, Work, Optimality, Flexibility, Stability...
- Individuals in the information age are motivated by more than commerce
  - Passion, Freedom, Social worth, Openness, Activity, Caring, Creativity
- "The Hacker Ethic and the Spirit of the Information Age"
  - Pekka Himanen & Linus Torvalds
- If companies clutch too tightly at short-term commercial success, they will kill partnerships, disenchant their employees and developers, and drive out innovation

Smartphones: open for innovation and creativity

symbian

#6: Hubris

- Trying to take too large a leap in one step
  - .... Underestimating the complexity of what you're attempting
  - .... Thinking you can manage something by your own power

- "Evolution, not creation"
  - .... The eventual version of massively innovative technology is almost always attained through a series of incremental steps
  - .... Where each individual steps adds some extra value
  - .... *And makes commercial sense in its own right*

- That's why we need a **credible roadmap**
  - .... Credible history of delivery and problem solving

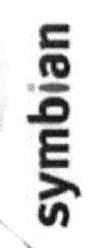

Example product roadmap

- Six-monthly releases aimed at all customers
- Designed in close collaboration with customers and partners
- Backward compatibility across all v9 releases – the bedrock for Ecosystem growth
- Market-leading security model in v9 (disruptive change)

First handset shipments

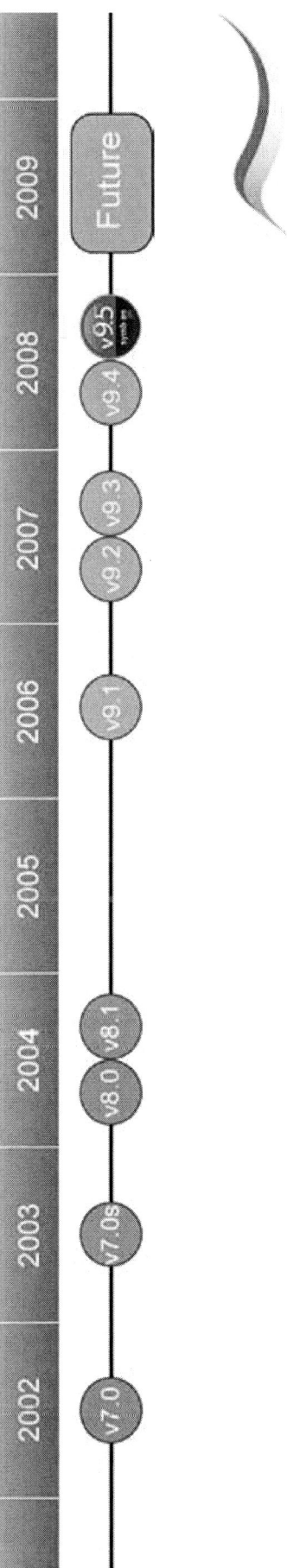

Achieving the historic potential of the smartphone

Smart phones = Smart lives

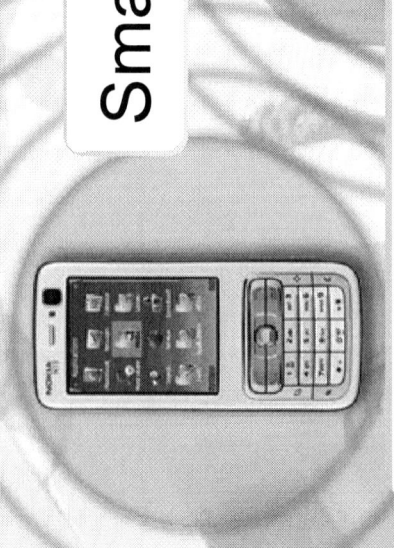

More achieved
More leisure
More pleasure…

- Rich functionality without excess power drain and cost
- Software & technology at the service of users, rather than overwhelming users
- Achieving both security and openness

Your freedom to fulfil *your* vision

- Services that work well, regardless of the choice of device, choice of network, or choice of manufacturer
- Motivation and fulfilment of the huge creative potential of individual smartphone users and contributors
- Regular steady improvements

symbian

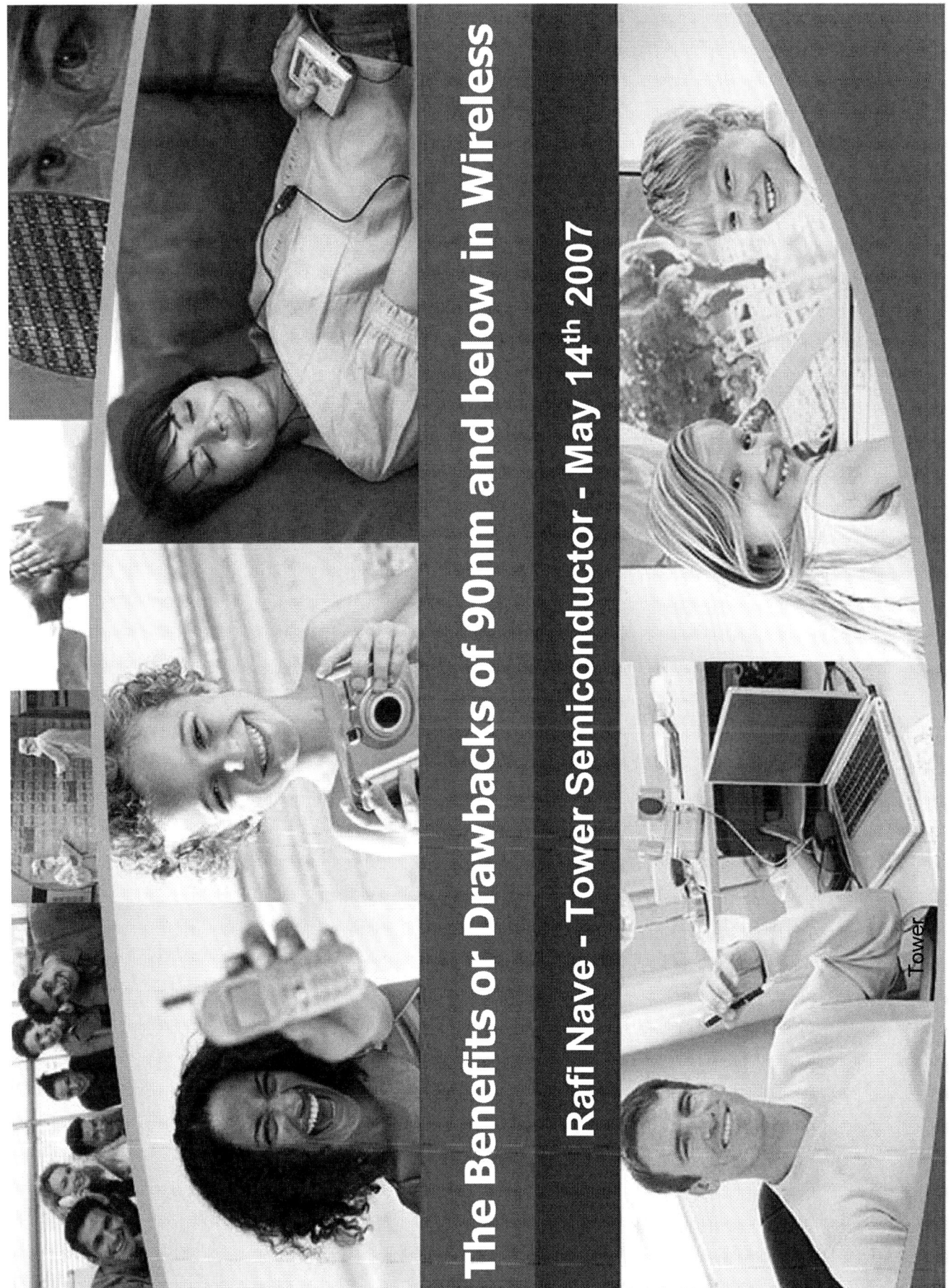

The Benefits or Drawbacks of 90nm and below in Wireless

Rafi Nave - Tower Semiconductor - May 14th 2007

The Benefits or Drawbacks of 90nm and below in Wireless

- The common myth
- The reality
- Tower's experience
- Conclusions

The common myth

- "Small is beautiful" has been been the drum to which the Semiconductor industry was marching for 4 decades
  - Moore's law
  - The perception that if you don't move ahead you stay behind (treadmill syndrome)

- The compaction to the latest & greatest process node was assumed to bring performance and cost benefits

- It is always 'in' to ride the front of the wave [Avant guarde]

The reality

- **The new technology generations are very costly!**
  - Product development cost sky-rockets

  - The 3D effects & others reduce 1st Si success rate significantly

  - Mask set cost and other NREs very costly

- **In many cases Wireless products don't benefit as much as digital chips**
  - The Analog portions (transmitter, receiver etc) don't scale and hence consume expensive chip size in the advanced node.

Tower's experience

- Tower developed RF, and specifically wireless, offering in its 0.18 micron process

- Several customers implemented impressive Data communication products in this 0.18u RF-CMOS node

  - Including 802.11 Wireless LAN products attaining frequency of 5.9 GigaHertz

- This performance is available at the cost-effective 0.18u process and without need of BiCMOS, SiGe or other costlier ingredients

- Feasibility of much higher frequencies, up to 24GHz has been proven

Conclusions

- Before you rush to the latest and greatest processes – check if you really need what they offer

- The decision should be driven by economical considerations coupled with the product performance and quality traits

Cost per new technology node

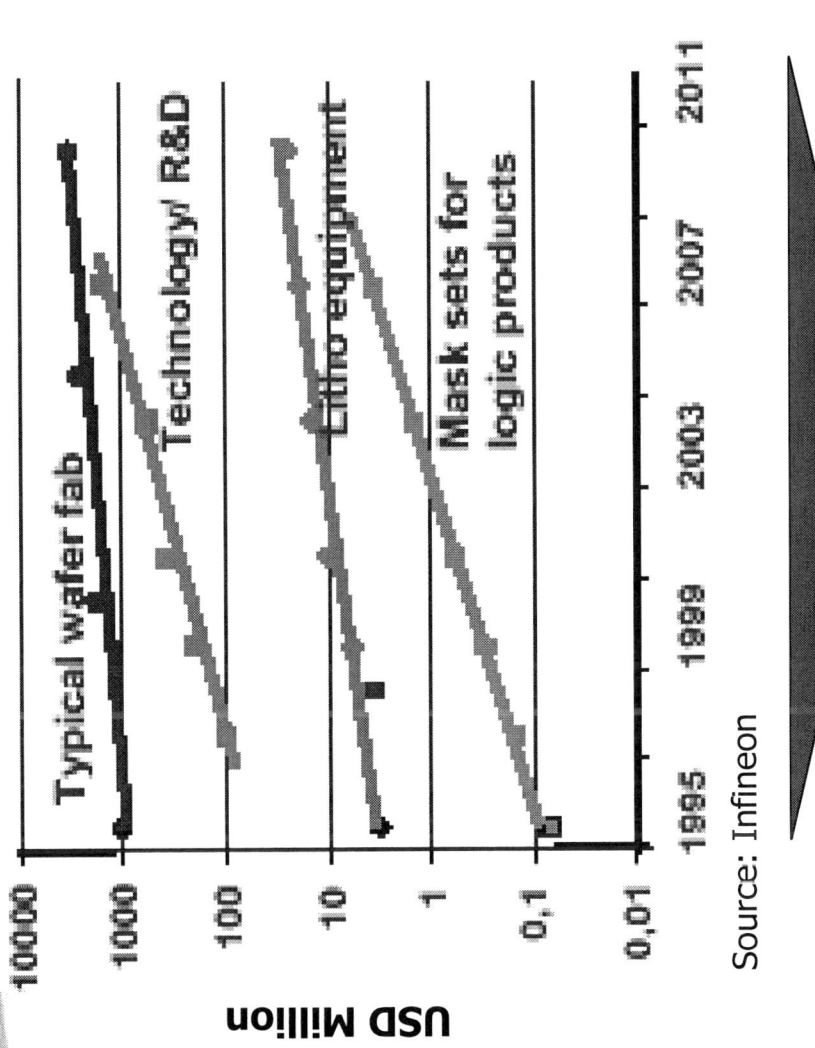

Source: Infineon

Cost per new technology node rises faster than the addressable market

Silicon success rates declining

□ *Survey-based silicon success rates declining*

- **First silicon OK: 48% in 2000 → 39% in 2002 → 34% in 2004**

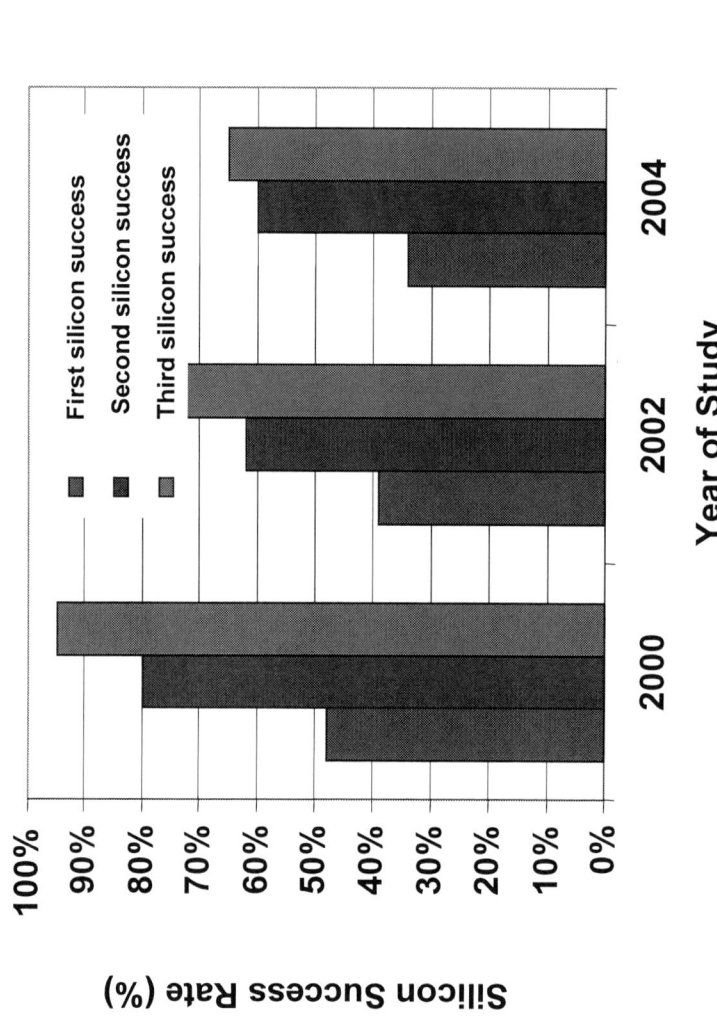

Collett International Research: 2000, 2002 Functional Verification Studies; 2003 Design Closure Study, 01/04

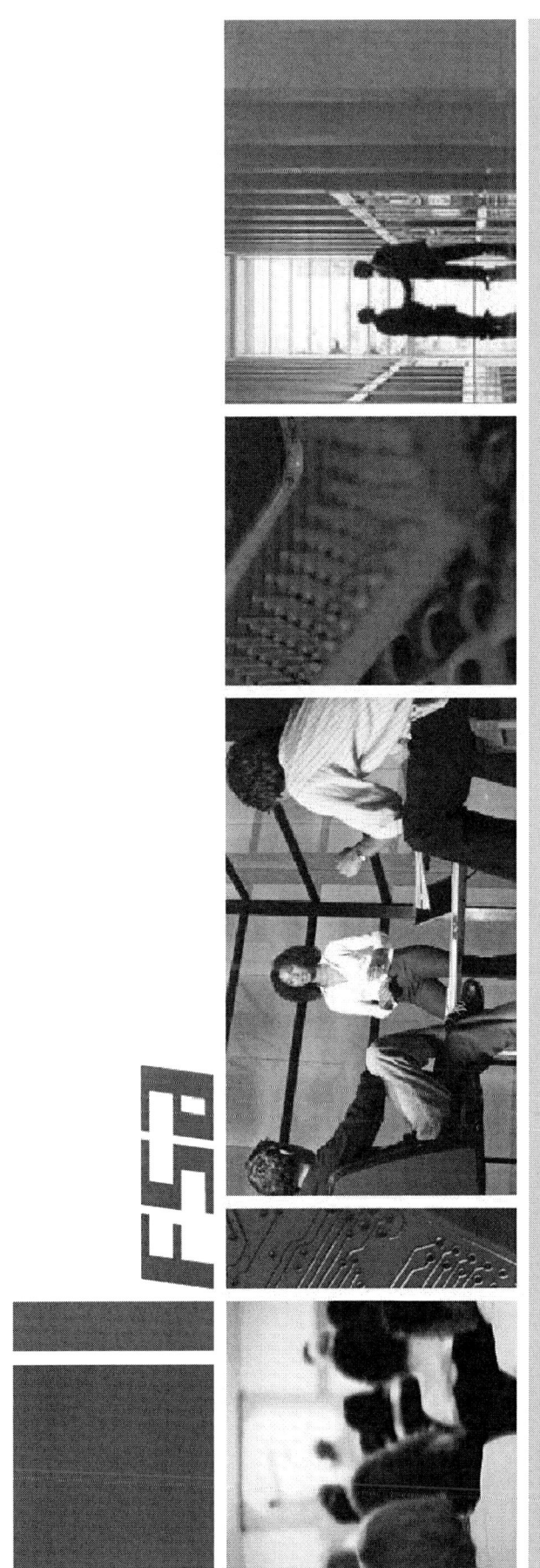

Consumer Electronics and the Semiconductor Industry
-- Perspectives for a Paradigm Shift

John Yu
COO, Chipnuts Technology Inc.
May 15,2007

AGENDA

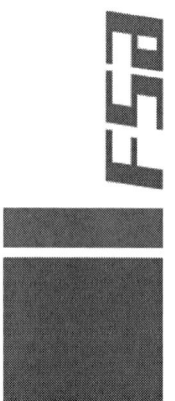

- Background
- Consumer electronics market characteristics
- China impact
- Challenges for semiconductor supply chain
- The road ahead

AGENDA

FSA

- **Background**
  - Consumer electronics market characteristics
  - China impact
  - Challenges for semiconductor supply chain
  - The road ahead

ELECTRONICS END MARKETS

- Dominated by data processing, communication, and consumer

Semi Revenue as a Percentage of Electronics

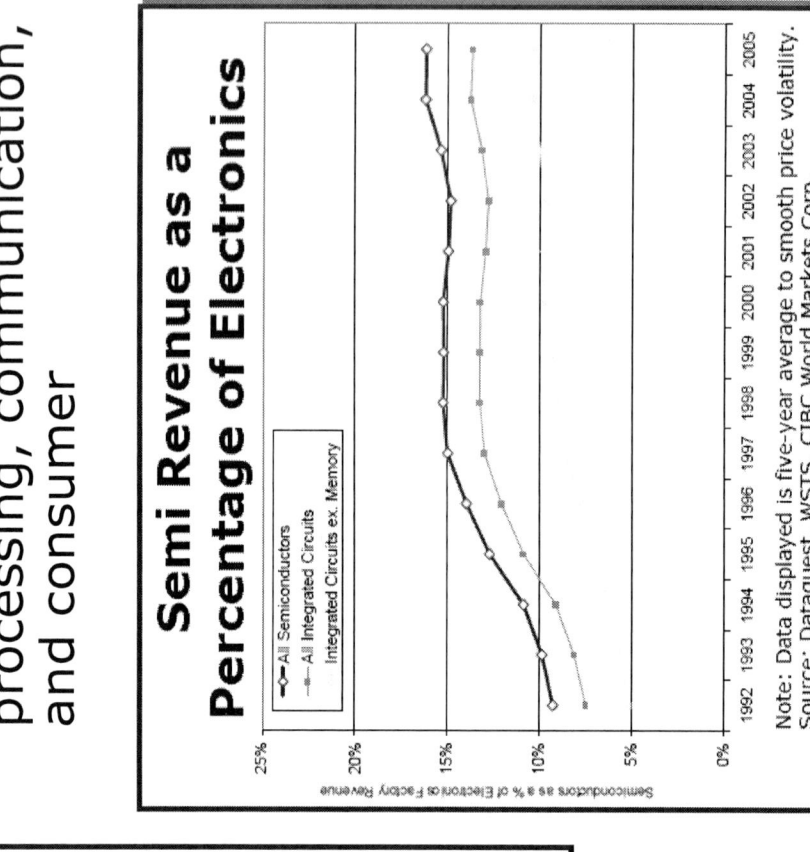

Semiconductors as a % of Electronics Factory Revenue

- All Semiconductors
- All Integrated Circuits
- Integrated Circuits ex. Memory

Note: Data displayed is five-year average to smooth price volatility.
Source: Dataquest, WSTS, CIBC World Markets Corp.

Electronics Revenue By End Market

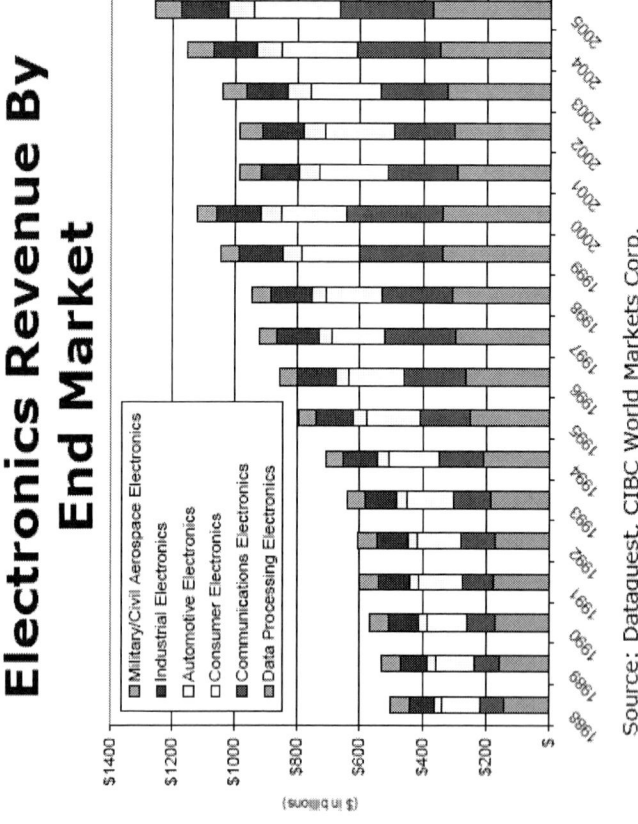

($ in billions)

- Military/Civil Aerospace Electronics
- Industrial Electronics
- Automotive Electronics
- Consumer Electronics
- Communications Electronics
- Data Processing Electronics

Source: Dataquest, CIBC World Markets Corp.

- Importance of semiconductor rising in end products

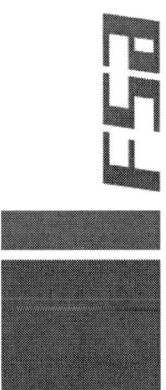

SEMICONDUCTOR CONSUMPTION

Worldwide Semiconductor Consumption by Region (M$)

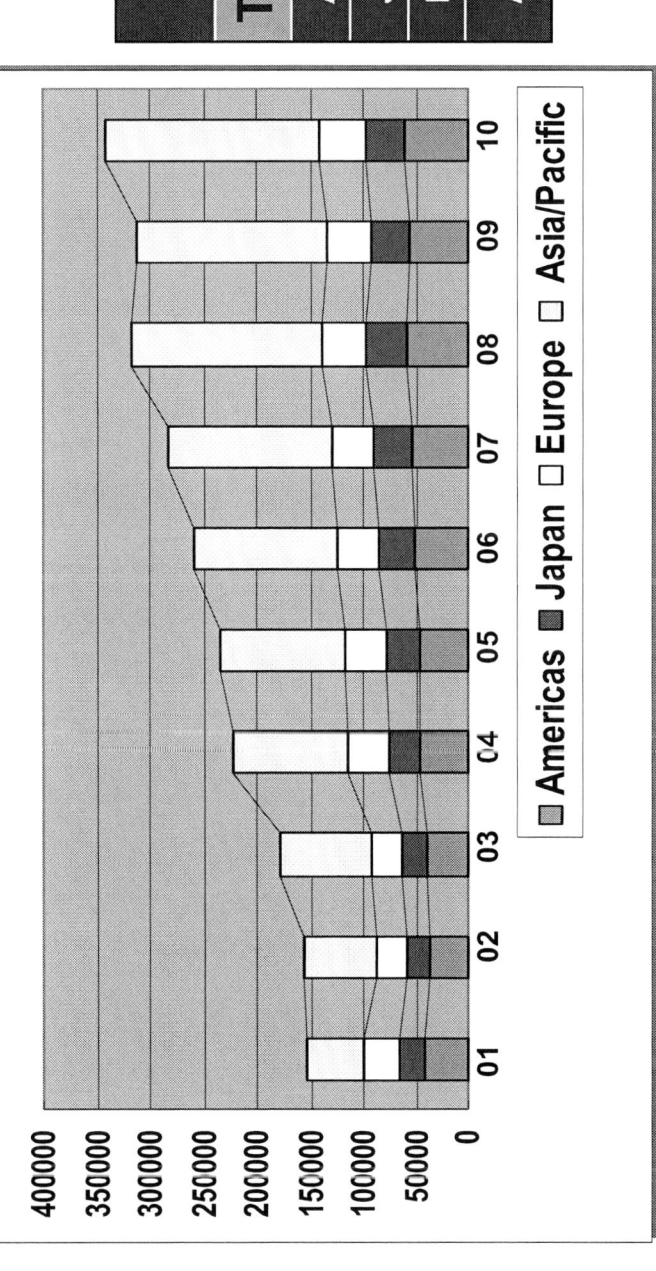

	CAGR 2004-2010
Total	7.4%
Americas	4.7%
Japan	3.2%
Europe	3.1%
Asia/Pacific	10.7%

Source: Gartner Dataquest (November 2006)

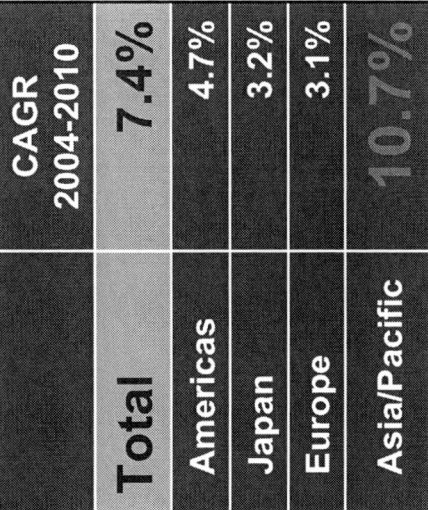

ASIC/ASSP CONSUMPTION BY APPLICATION

Source: Gartner Dataquest (November 2006)

Worldwide ASIC Consumption (M$)

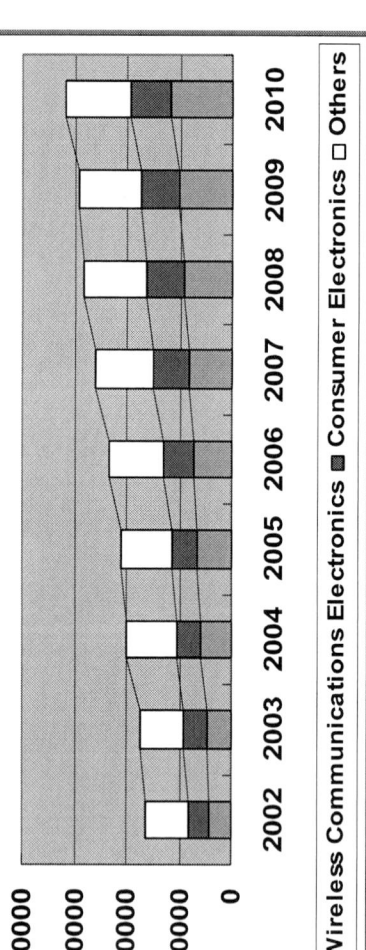

2002 2003 2004 2005 2006 2007 2008 2009 2010

40000 30000 20000 10000 0

■ Wireless Communications Electronics ■ Consumer Electronics □ Others

ASIC	CAGR 2004-2010
Wireless Comm.	12.1%
CE	8.9%
Others	N/A
Total	**8.1%**

Worldwide ASSP Consumption (M$)

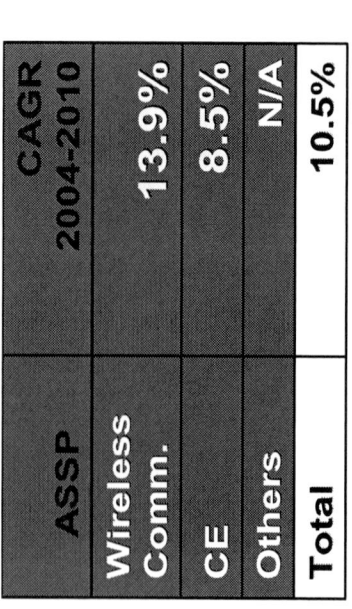

2002 2003 2004 2005 2006 2007 2008 2009 2010

100000 80000 60000 40000 20000 0

■ Wireless Communications Electronics ■ Consumer Electronics □ Others

ASSP	CAGR 2004-2010
Wireless Comm.	13.9%
CE	8.5%
Others	N/A
Total	**10.5%**

ASIC/ASSP CONSUMPTION BY REGION

Source: Gartner Dataquest (November 2006)

ASIC	CAGR 2004-2010
Americas	4.6%
Japan	1.6%
Europe	2.6%
Asia-Pacific	13.1%

ASIC Consumption Growth by Region

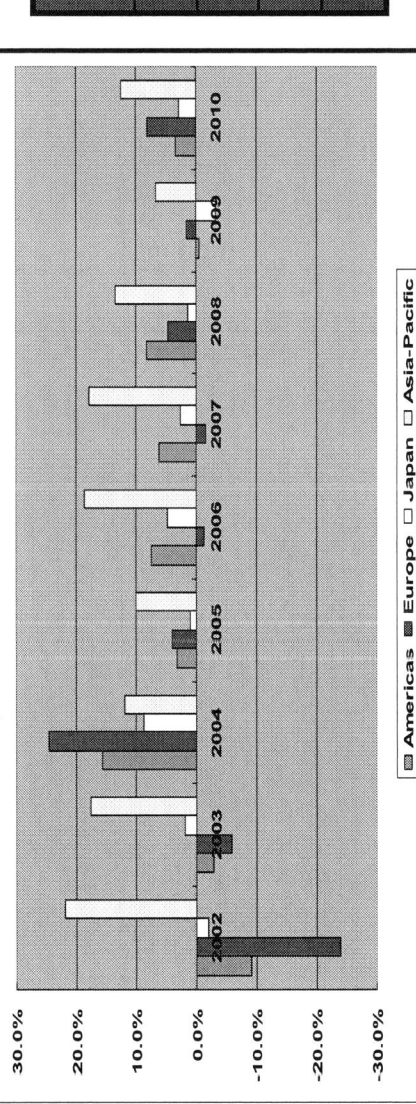

Americas ■ Europe □ Japan □ Asia-Pacific

ASSP Consumption Growth by Region

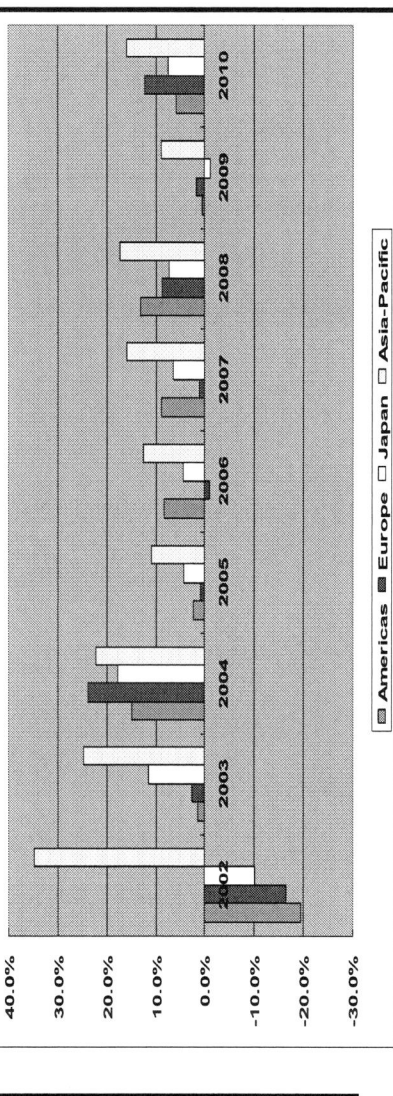

Americas ■ Europe □ Japan □ Asia-Pacific

ASSP	CAGR 2004-2010
Americas	6.4%
Japan	4.7%
Europe	3.8%
Asia-Pacific	13.6%

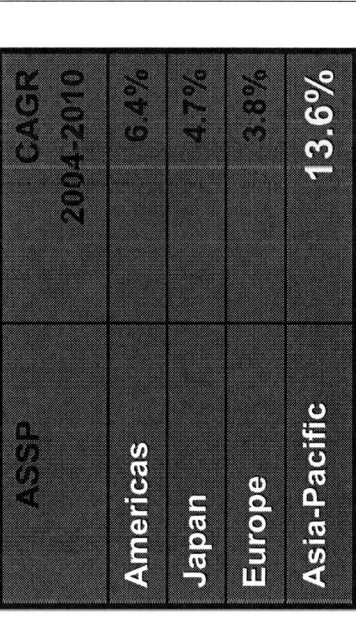

PROCESS NODES

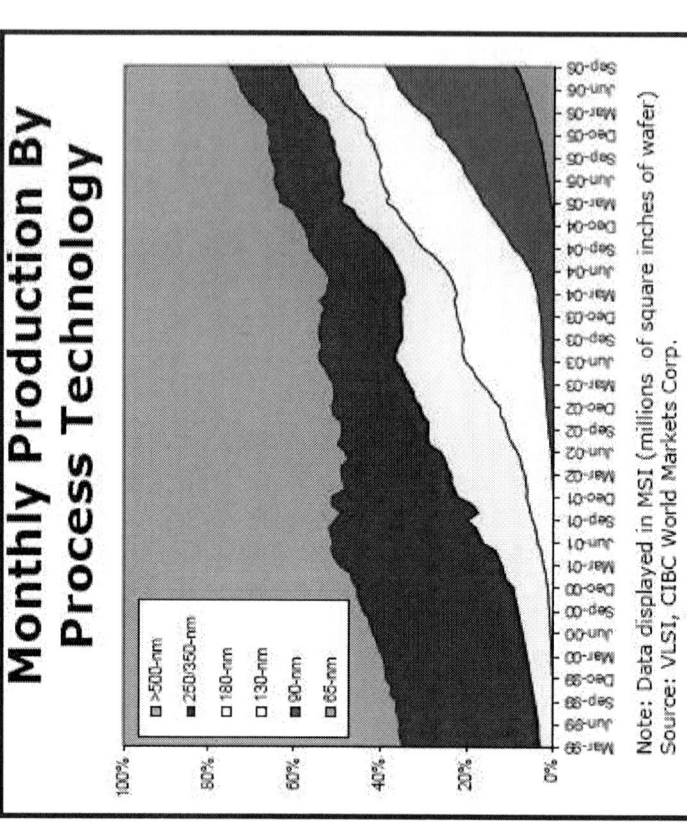

Monthly Production By Process Technology

Note: Data displayed in MSI (millions of square inches of wafer)
Source: VLSI, CIBC World Markets Corp.

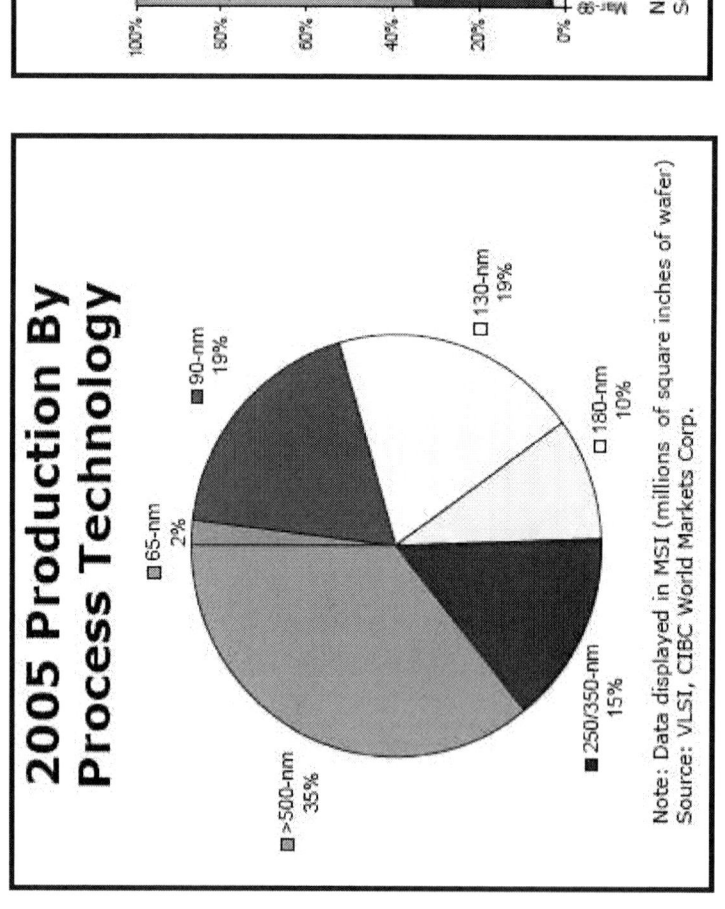

2005 Production By Process Technology

Note: Data displayed in MSI (millions of square inches of wafer)
Source: VLSI, CIBC World Markets Corp.

* 45-nm products are expected to be commercially available in mid 2007 to early 2008.

AGENDA

FSA

- Background
- **Consumer electronics market characteristics**
- China impact
- Challenges for semiconductor supply chain
- The road ahead

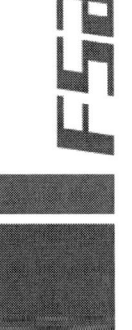

CONSUMER MARKET CHARACTERISTICS

- Very short product life

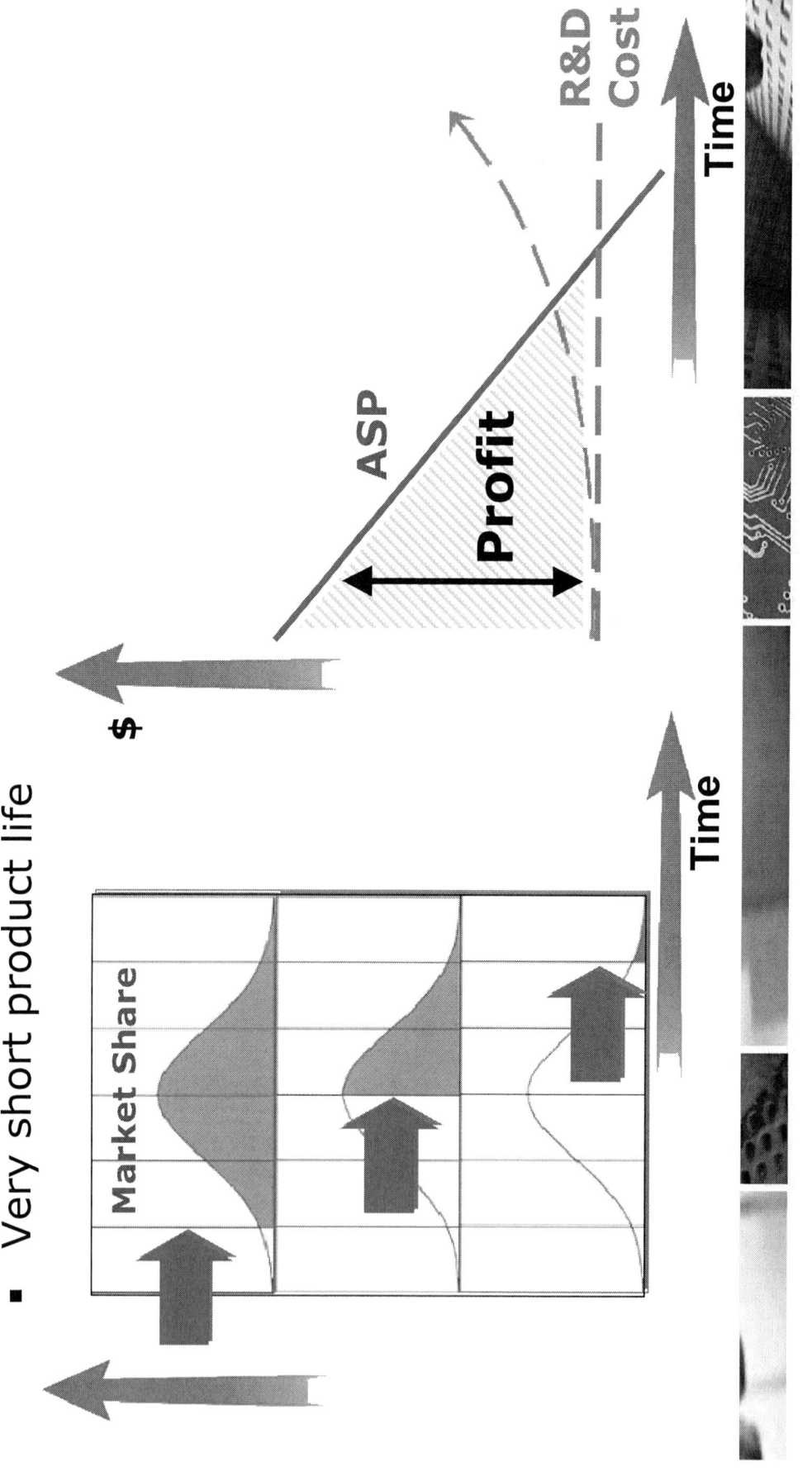

FSA CONSUMER MARKET CHARACTERISTICS

- Very short product life
- Rapid feature enhancement

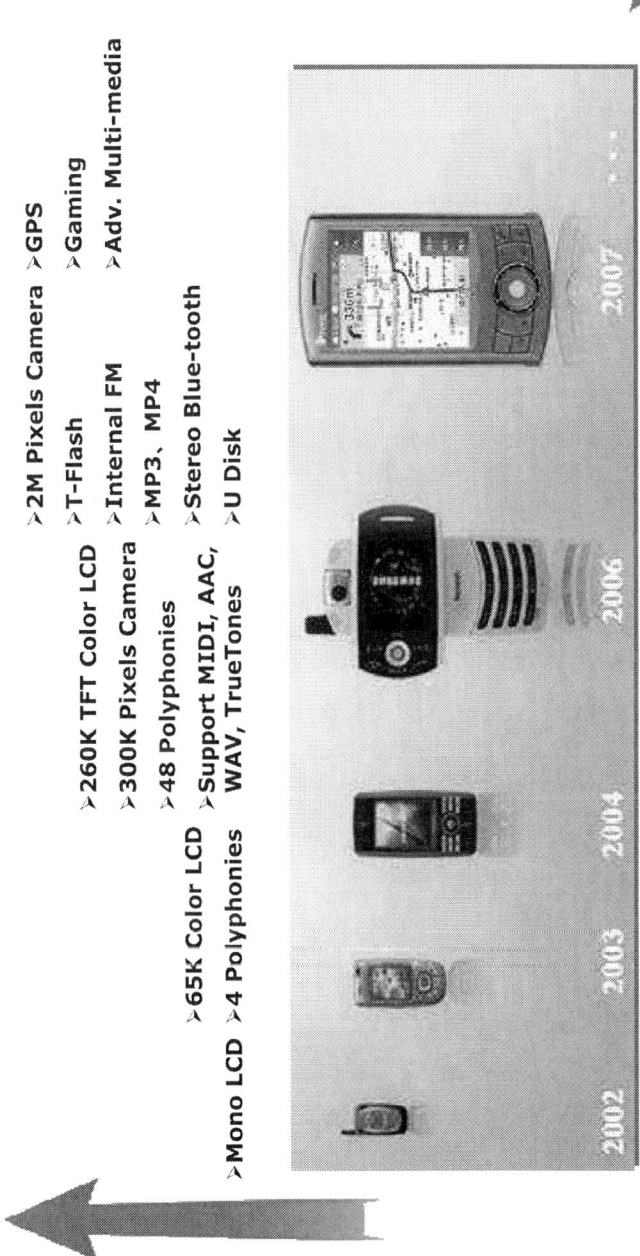

> Mono LCD > 4 Polyphonies

> 65K Color LCD

> 260K TFT Color LCD
> 300K Pixels Camera
> 48 Polyphonies

> 2M Pixels Camera > GPS
> T-Flash > Gaming
> Internal FM > Adv. Multi-media
> MP3、MP4
> Stereo Blue-tooth
> U Disk

> Support MIDI, AAC,
WAV, TrueTones

2002   2003   2004   2006   2007

CONSUMER MARKET CHARACTERISTICS

- Very short product life
- Rapid feature enhancement
- Personalization

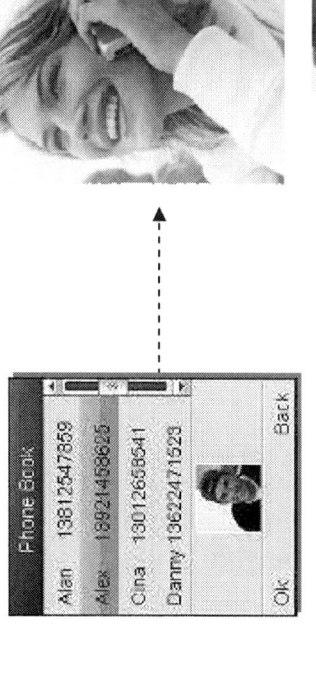
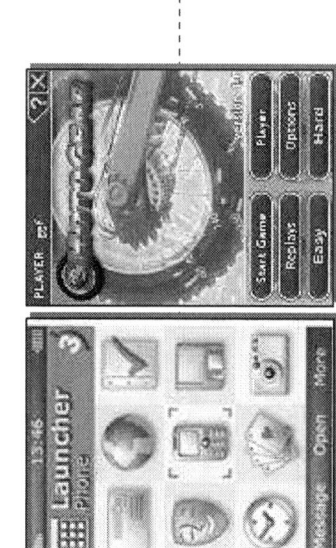
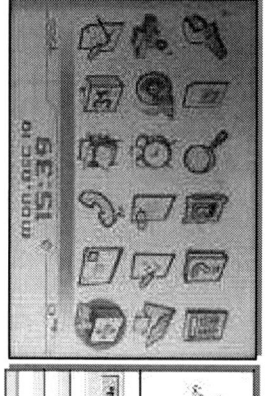
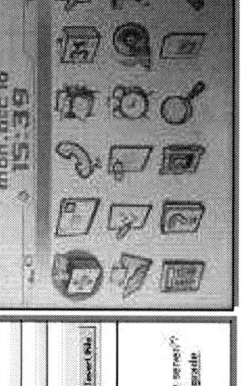

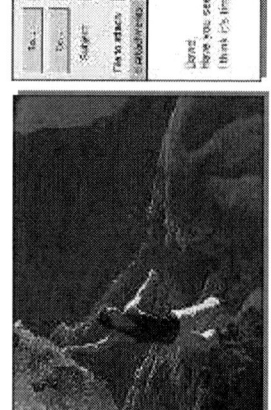

CONSUMER MARKET CHARACTERISTICS

- Very short product life
- Rapid feature enhancement
- Personalization
- **Ease of Use**

CONSUMER MARKET CHARACTERISTICS

- Very short product life
- Rapid feature enhancement
- Personalization
- Ease of Use
- **Affordability**

Volume

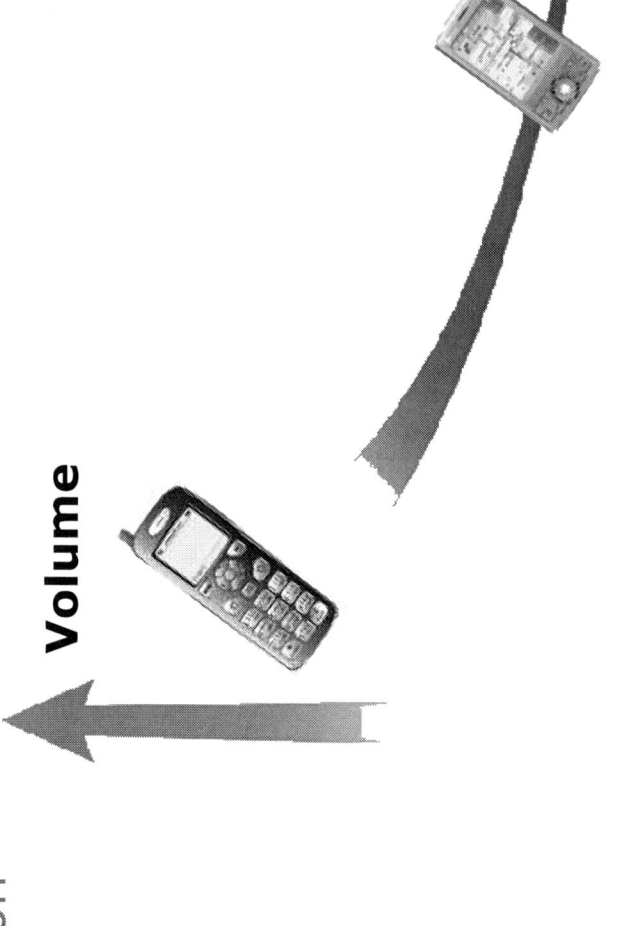

CONSUMER MARKET CHARACTERISTICS

- Extremely short product life
- Rapid feature enhancement
- Personalization
- Ease of use
- Affordability
- Fast ramp up/down

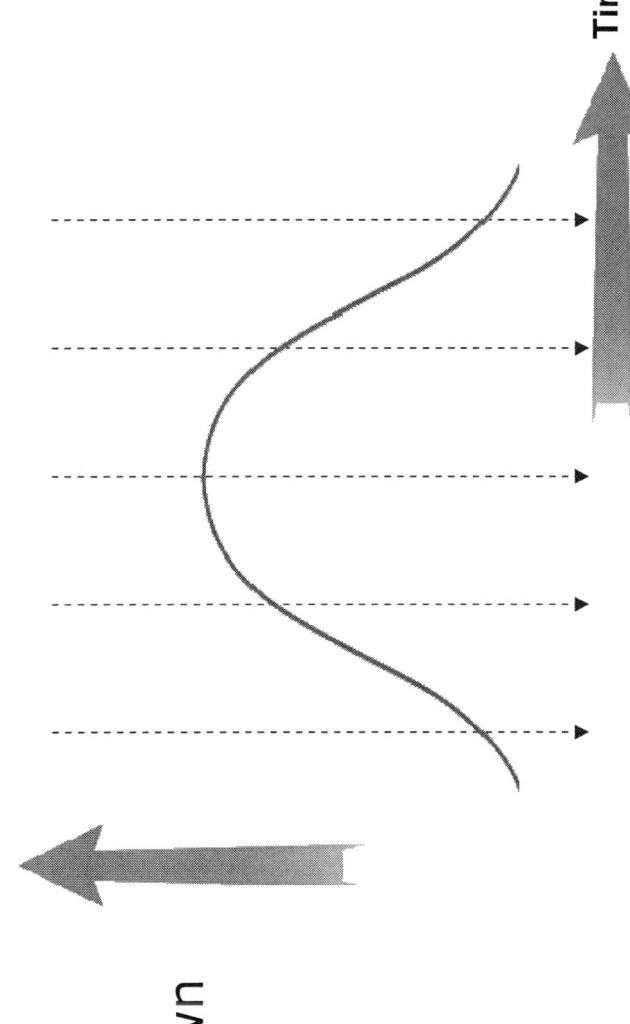

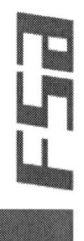

AGENDA

- Background
- Consumer market characteristics
- **China impact**
- Challenges for semiconductor supply chain
- The road ahead

CHINA IC CONSUMPTION

IC consumption growth in China market (M$)

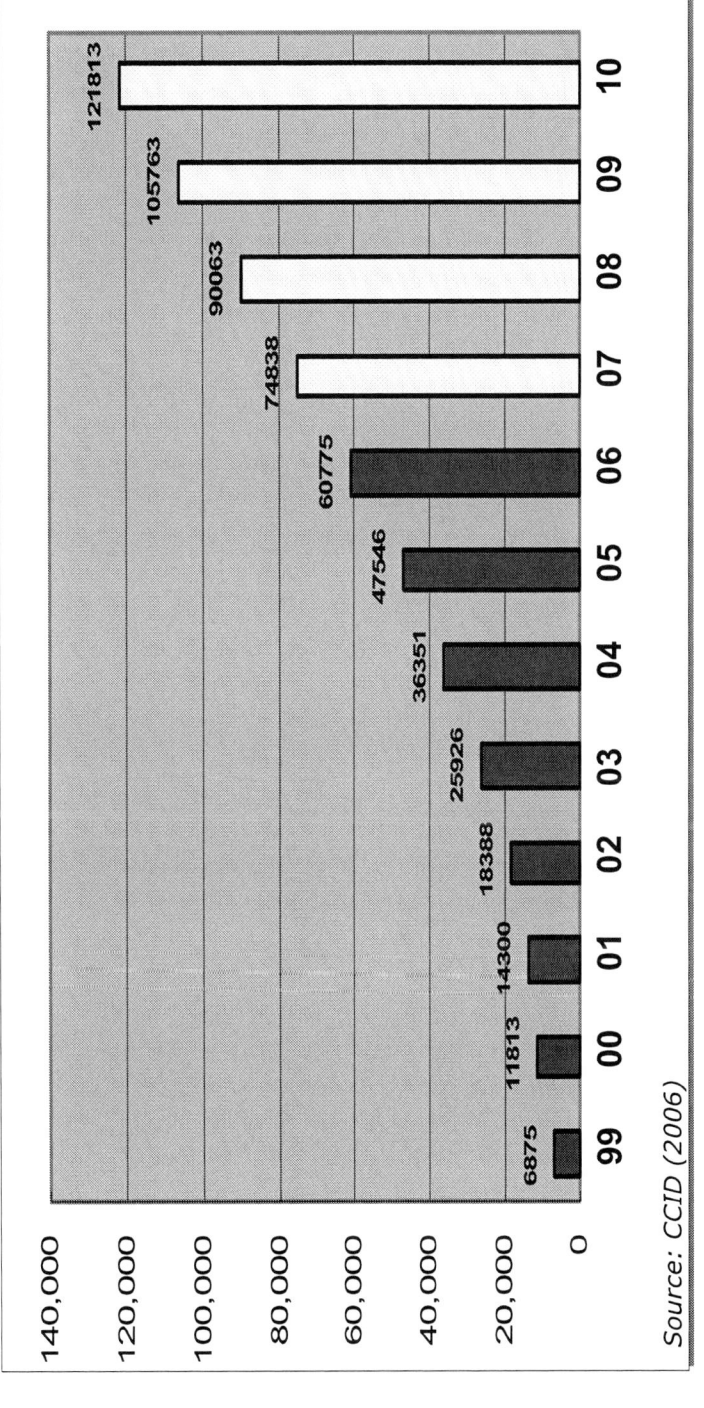

Year	Value
99	6875
00	11813
01	14300
02	18388
03	25926
04	36351
05	47546
06	60775
07	74838
08	90063
09	105763
10	121813

Source: CCID (2006)

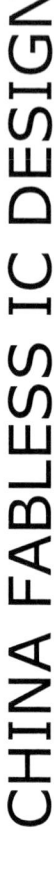

CHINA FABLESS IC DESIGN

China Fabless Design Houses

Year	Value
2001	243
2002	390
2003	463
2004	476
2005	479
2006	450

Source: CSIA

Only **10%** of fabless design companies have products in MP

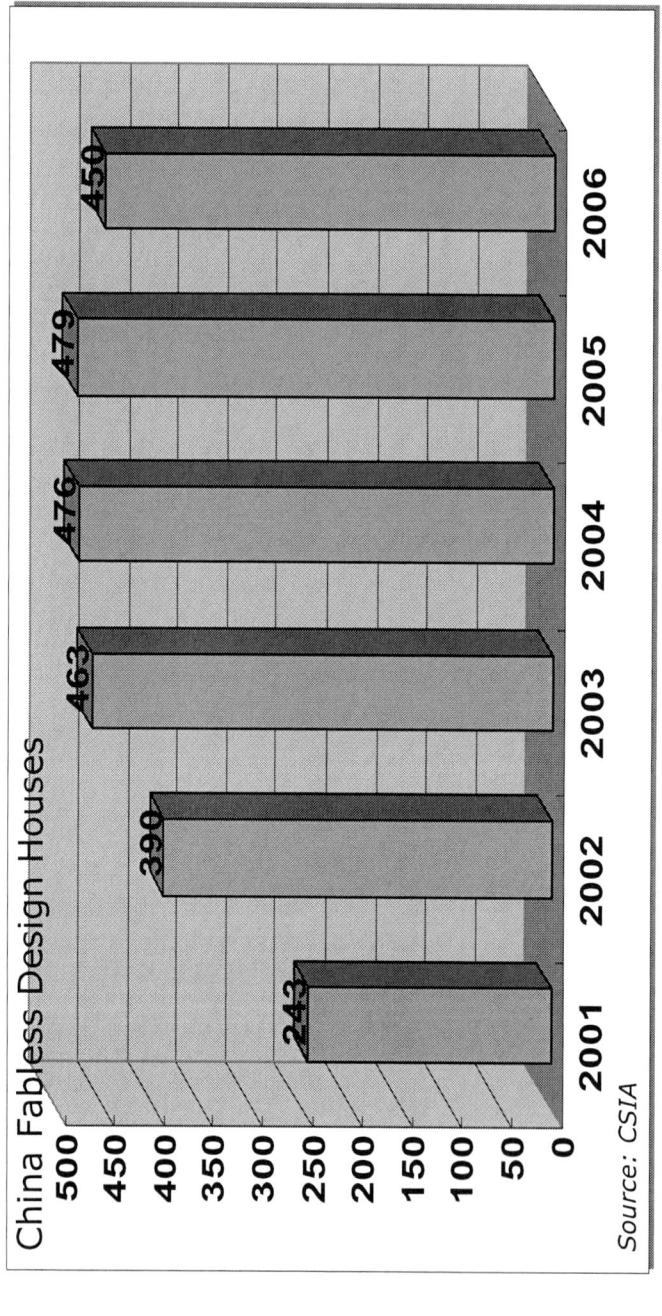

MARKET IMPACT

- Huge population → huge potential customers

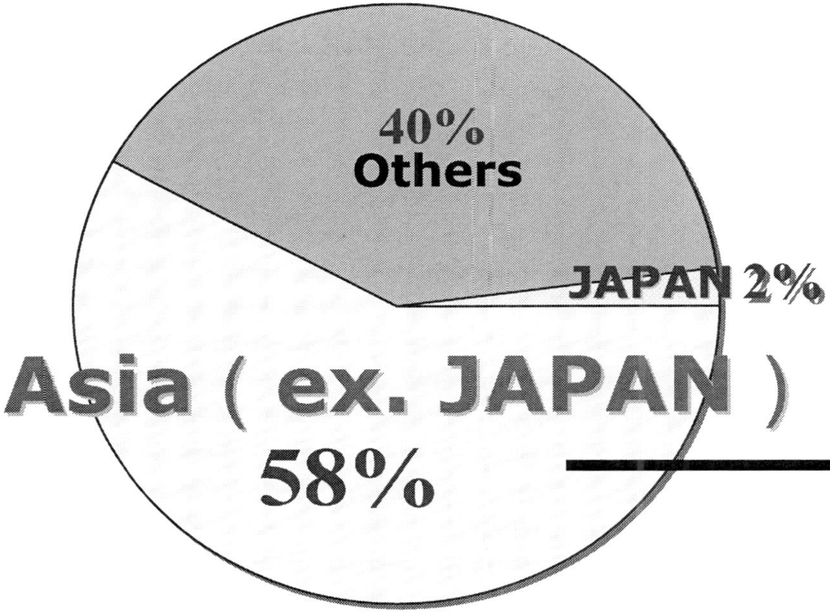

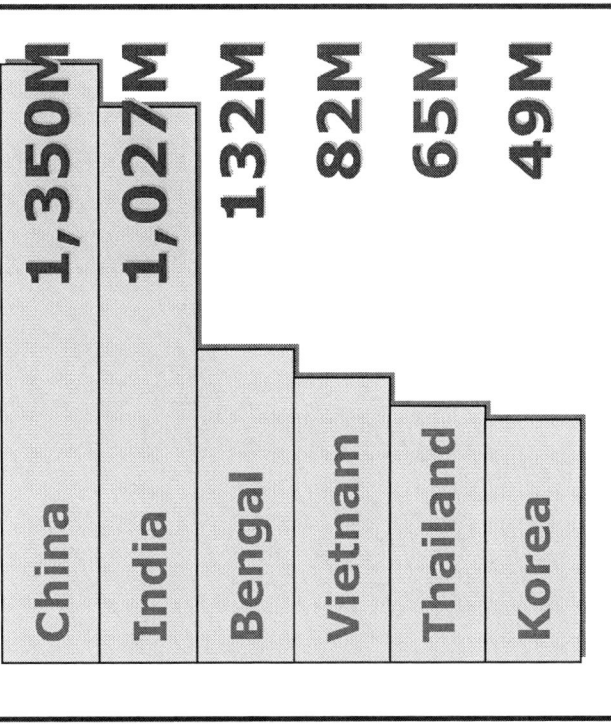

China	1,350M
India	1,027M
Bengal	132M
Vietnam	82M
Thailand	65M
Korea	49M

Asia (ex. JAPAN)
58%

Others 40%

JAPAN 2%

Mobile phone penetration < 30%

MARKET IMPACT

- Huge population → huge potential customers
- Wide affordability range → support various price points

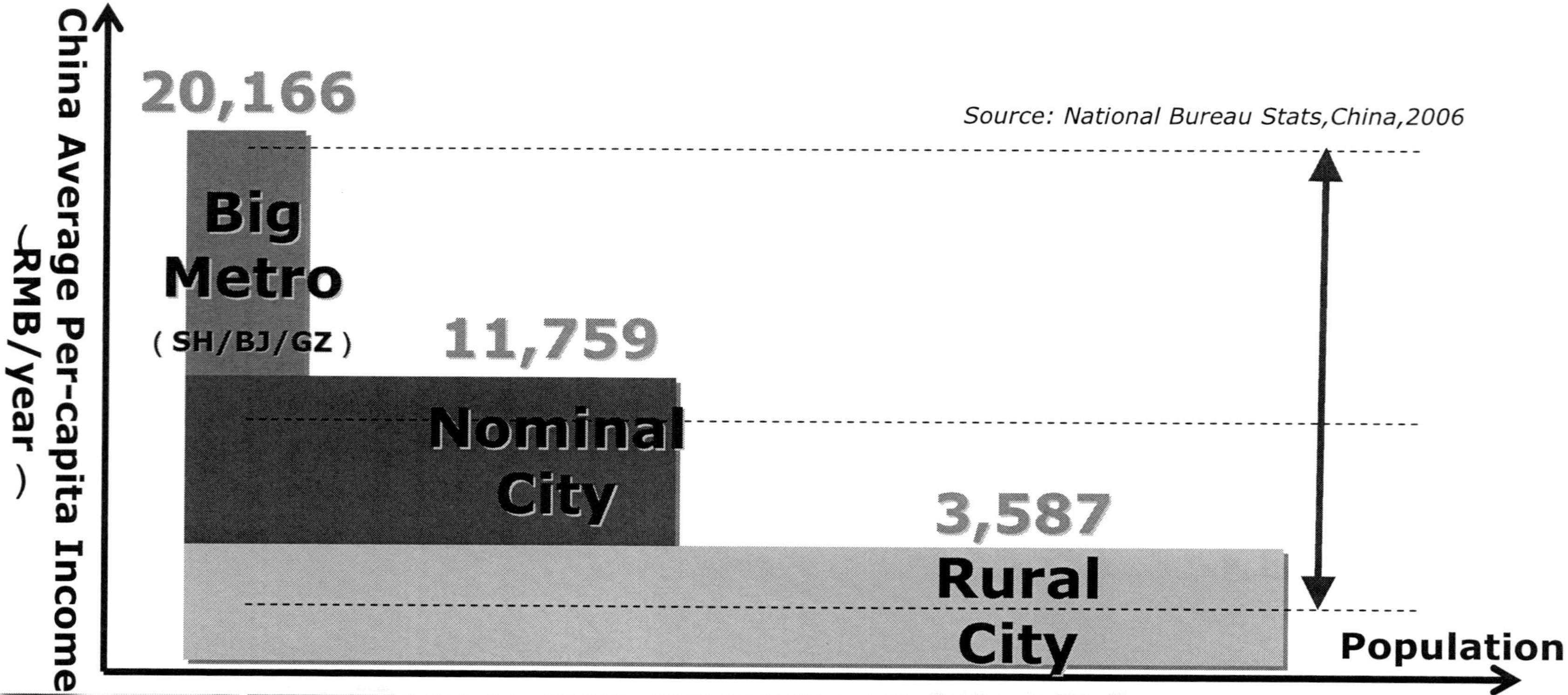

Source: National Bureau Stats,China,2006

Population

20,166

Big Metro
(SH / BJ / GZ)

11,759

Nominal City

3,587

Rural City

China Average Per-capita Income
⁓**RMB/year**⁓

MARKET IMPACT

China Handset Market Price Distribution

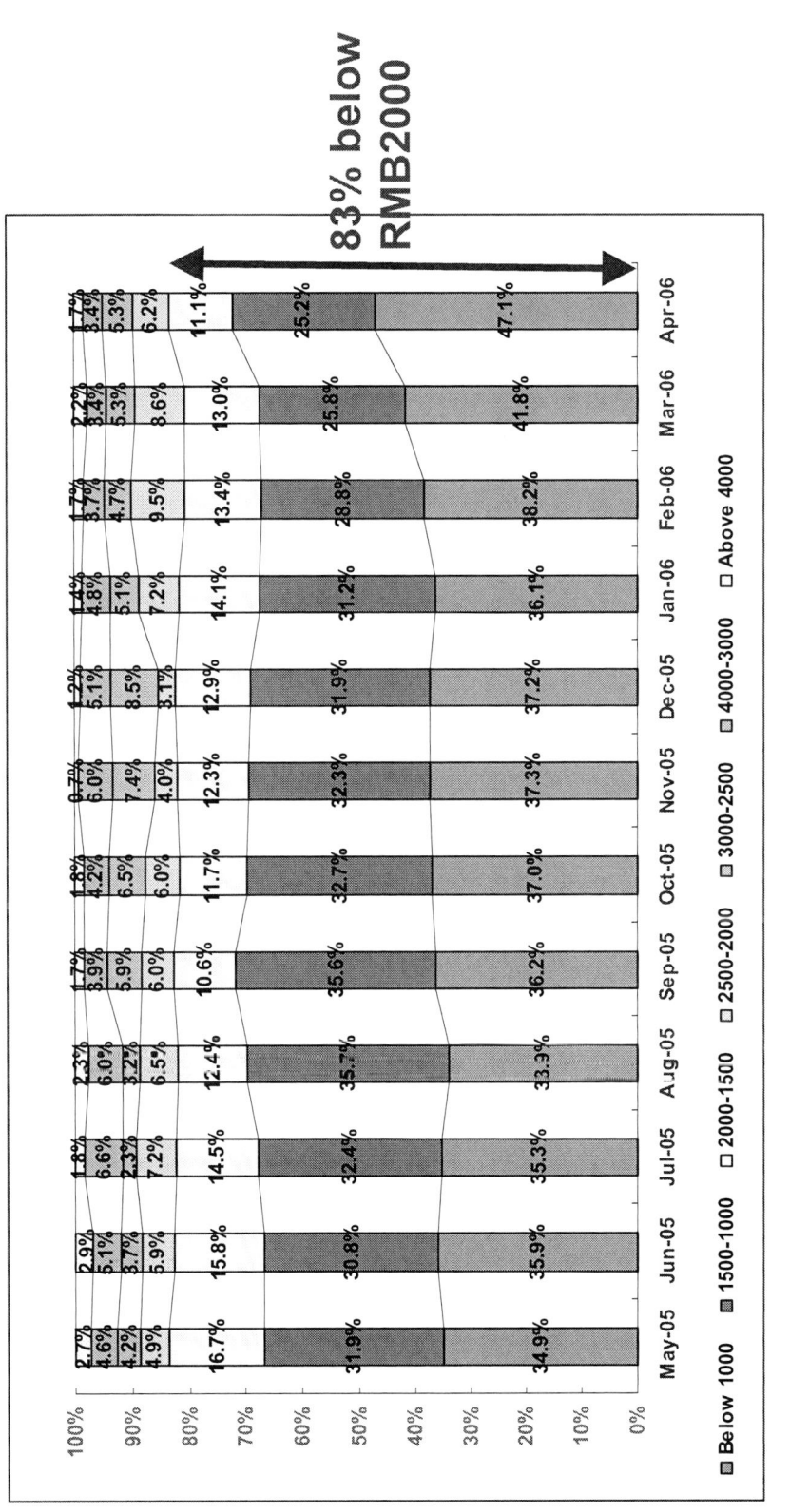

83% below RMB2000

MARKET IMPACT

- Huge population → huge potential customers
- Wide affordability range → support various price points
- **Very fast replace cycles → faster feature update, shorter time-to-market required**

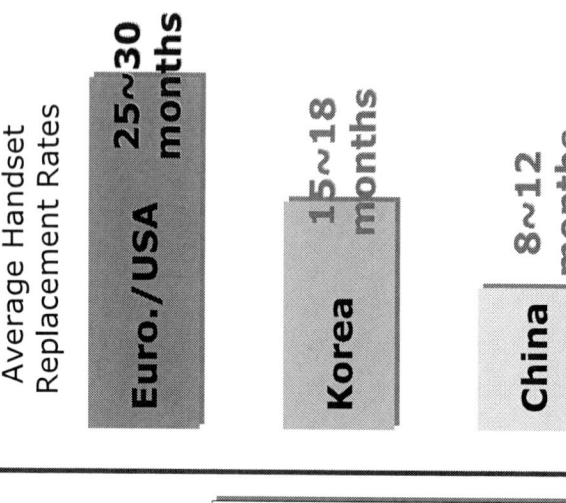

Average Handset Replacement Rates

Euro./USA 25~30 months

Korea 15~18 months

China 8~12 months

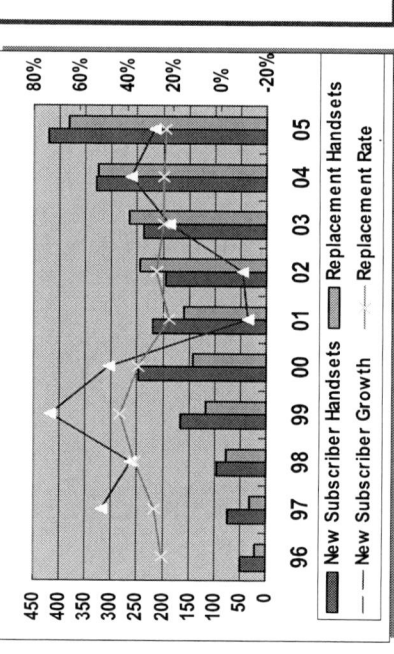

Worldwide Replacement vs. New Sub (M)

New Subscriber Handsets — Replacement Handsets — New Subscriber Growth — Replacement Rate

Worldwide Handset Unit Shipments (M)

Handset Unit Shipments — YoY Growth

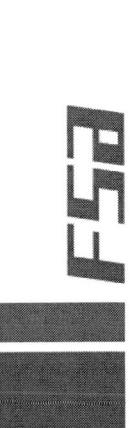

SUPPLY CHAIN IMPACT

- Huge, efficient manufacturing base

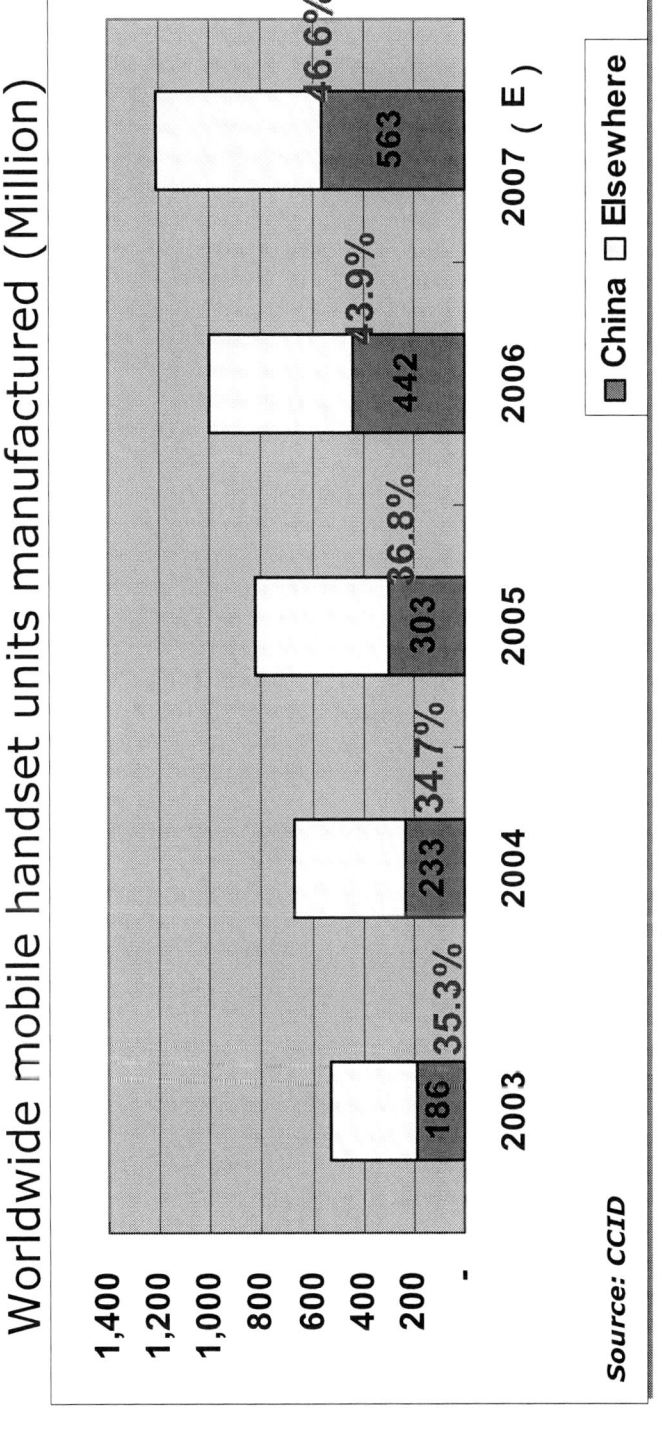

Worldwide mobile handset units manufactured (Million)

	2003	2004	2005	2006	2007 (E)
China	186	233	303	442	563
%	35.3%	34.7%	36.8%	43.9%	46.6%

■ China □ Elsewhere

Source: CCID

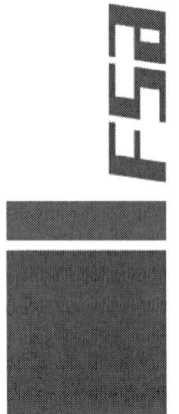

SUPPLY CHAIN IMPACT

- Huge, efficient manufacturing base
- Large labor pool that is ready to work

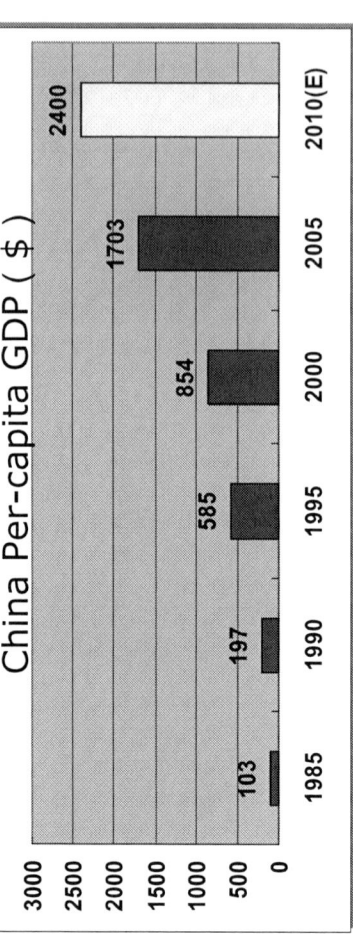

China Per-capita GDP ($)

Year	Value
1985	103
1990	197
1995	585
2000	854
2005	1703
2010(E)	2400

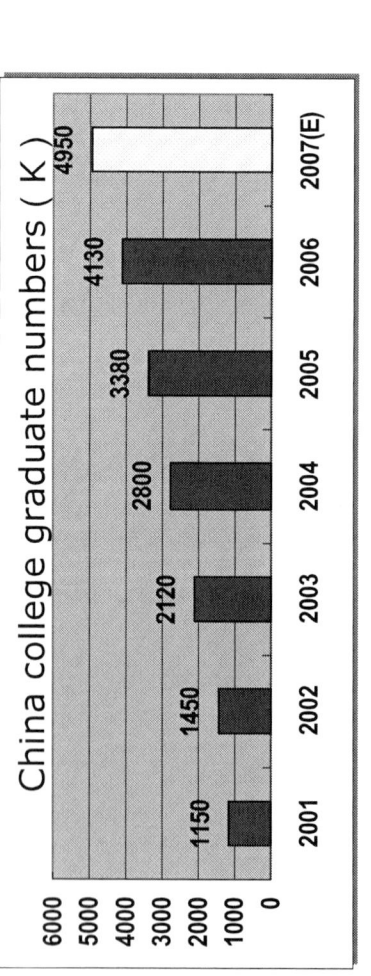

China college graduate numbers (K)

Year	Value
2001	1150
2002	1450
2003	2120
2004	2800
2005	3380
2006	4130
2007(E)	4950

Source: National Bureau Stats, China

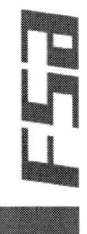

SUPPLY CHAIN IMPACT

- Huge, efficient manufacturing base
- Large labor pool that is ready to work
- Order of priority: Cost → Time → Performance

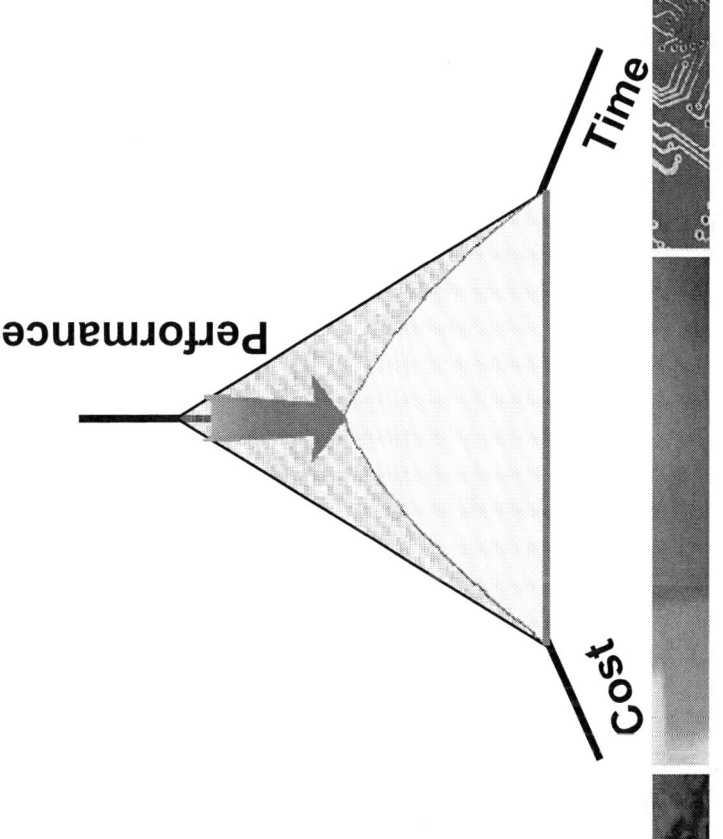

AGENDA

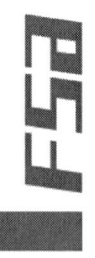

- Background
- Consumer market characteristics
- China impact
- **Challenges for semiconductor supply chain**
- The road ahead

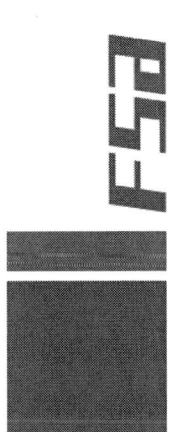

LEADING THEMES

- Speed
  - Time-to-market / volume
  - More demanding in China
- Cost at the right compromise of performance
  - Right cost to get "in the door"
  - Cost > features > performance
- Total solution
  - Shrink-wrapped and ready to go
  - Side-by-side support is a must
- Open and flexible
  - Multi-standards, multi-applications, multi-personalities!
  - Hardware-software trade-offs (hint: software is becoming more expensive!)

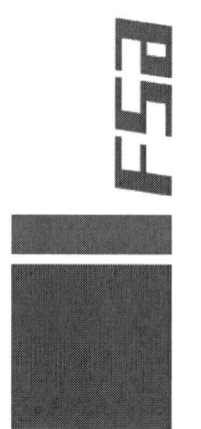

CHIP PLANNING AND SPECIFICATON

- Market-driven application & system specifications
- Application & system-driven chip specifications
- Hardware trade-offs in favor of software expandability
- Platform focused

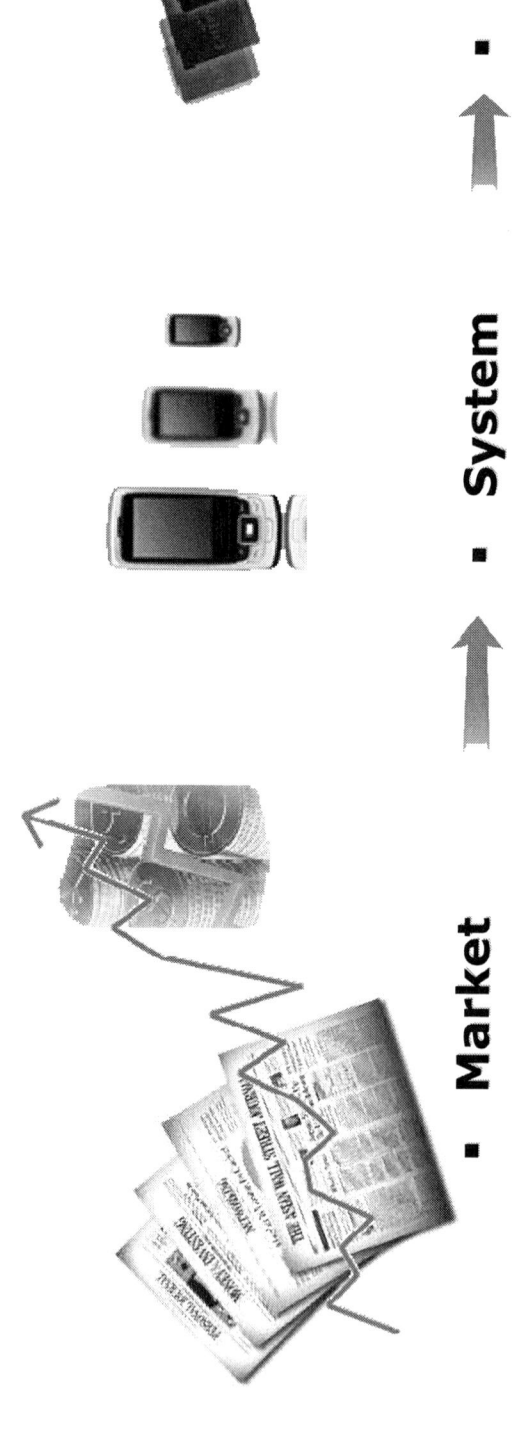

- **Market** → **System** → **SoC**

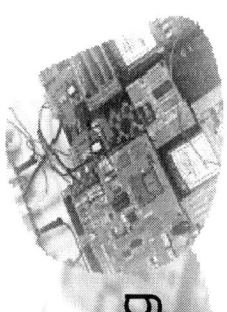

CHIP DESIGN

- System-chip co-design
- Hardware-software co-design
  - Early platform for software to start
- Co-verification on common platform
- System-wide cost, power, signal integrity management
- Cost focused
  - ... but leave margin for possible add-on features
  - Optimize for system, not just chip
  - Maximize leverage and re-use
- Rigorous program management
- Efficient abstracting with accurate modeling
- Defined quality objectives

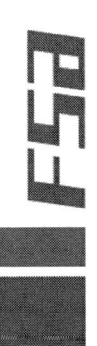

INTELLECTUAL PROPERTY

- Total solution
  - Not just an IP, but...
  - As enabler for complete solution (e.g. tools, docs, development platform, software, etc.)
- Chip and even system integration support
  - Made for re-use
- Robust and silicon-proven
  - Migratable
  - Shrinkable
- Flexible and creative business models

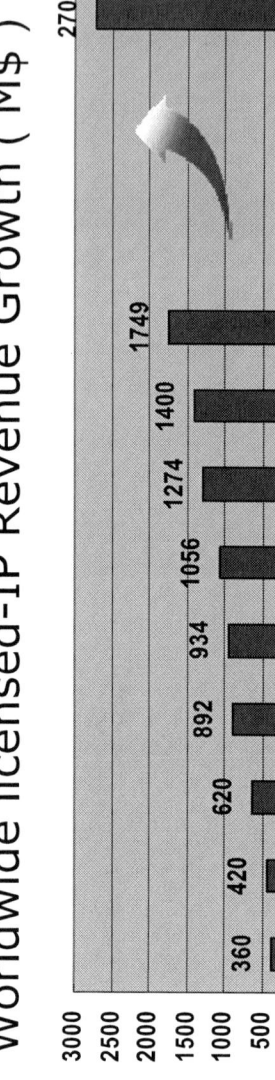

Worldwide licensed-IP Revenue Growth (M$)

360, 420, 620, 892, 934, 1056, 1274, 1400, 1749, 2700

1998, 1999, 2000, 2001, 2002, 2003, 2004, 2005, 2006, 2007, 2008, 2009, 2010(E)

3000, 2500, 2000, 1500, 1000, 500, 0

Source: Gartner

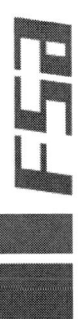

DESIGN METHODOLOGY AND TOOLS

- ESL tools
- Platform for both hardware and software development
- Production proven
- Widely supported by foundries and IP providers
- Enable re-usability and efficient migration
- Appropriate trade-off between accuracy and abstraction
- Easy to use, fast learning curve
- Integrated

MANUFACTURING

- Process technology criteria
  - High yield, tightly controlled window
  - Economy of scale
  - Wide selection of proven IPs
  - Easy and proven migration and cost-reduction paths
  - Able to ramp fast (capacity and fast cycle time)
- Packaging and assembly criteria
  - Variety for applications and cost levels, including SiP
  - Environmentally clean materials
  - Reliable and repeatable assembly
  - Able to ramp fast

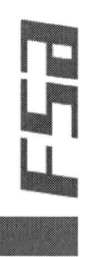

PRODUCT AND SOLUTION SUPPORT

- Total solution PDK
  - Production-ready HW and SW, SDK
  - Integrated and proven 3rd-party interfaces
  - Easy to use, fast learning curve
  - Modularization
  - Flexibility and stability
  - Compatibility and adoptability
  - Add-on cost minimization
  - Investment retention
  - Eco-systems
  - Timely and expert support

AGENDA

- Background
- Consumer market characteristics
- China impact
- Challenges for semiconductor supply chain
- **The road ahead**

CUSTOMERS WANT MORE FOR LESS

MORE	LESS
Features	Money
Quality and reliability	Time-to-volume
Flexibility, customizability and extensibility	Development time and add-on investment
Content support	User complexity
Integration	Power consumption

Supply chain must work as partners

NEW PERSPECTIVE OF "SoC"

FSA

- From System-on-Chip...

NEW PERSPECTIVE OF "SoC"

- From System-on-Chip... to "Solution-on-Chip"!

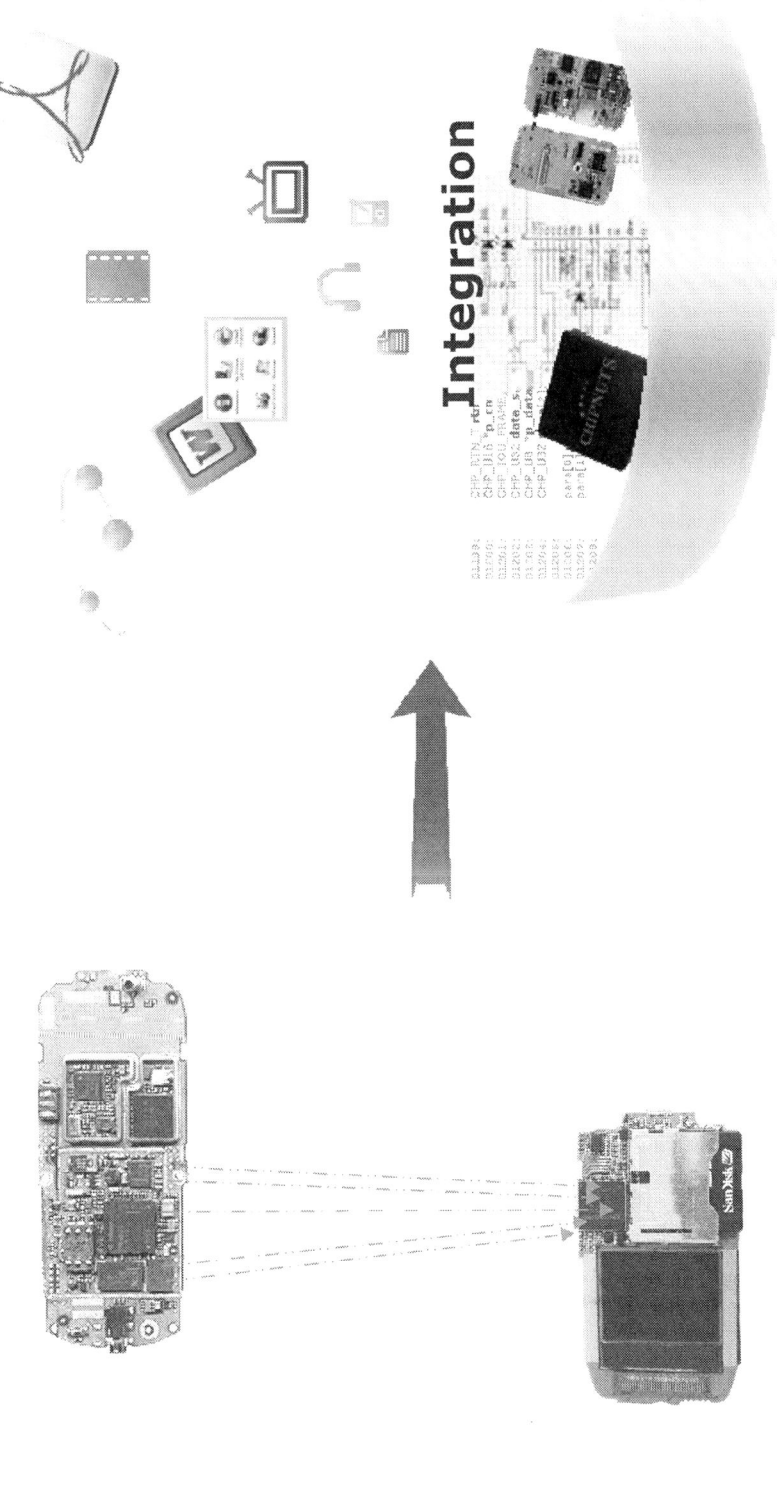

Integration

SHIFTING BOUNDARIES

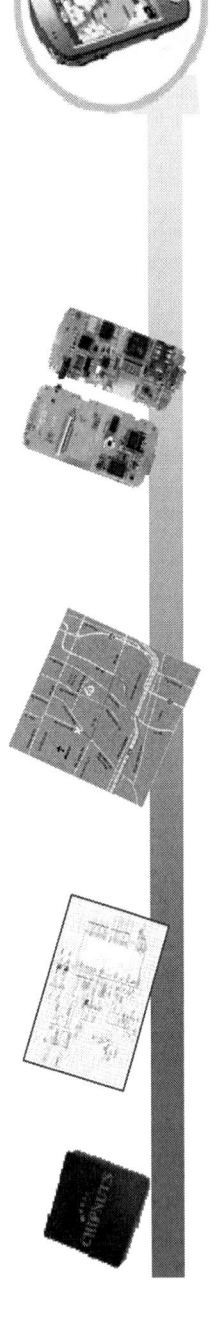

	Chip	System integration	Solution integration	Manufacturing	Product packaging	Sales & distribution
Today	Chipset vendor	System design house		EMS/OEM	Brand maker Channel distributor	
Future	Chipset vendor	System design house	EMS/OEM		Brand maker Channel distributor	

SUMMARY

F5B

- Consumer electronics market
  - Time-to-market
  - Affordable
  - Rich features that are customizable
  - Easy to use
  - More for less
- China impact
  - Huge opportunity and high growth rate
  - Large local manufacturing base
  - Relentless demand on speed to market, feature upgrades, and cost, especially cost
  - Much more for much less!
- Total solution offering for higher value and success
  - Mindset shift for each player in supply chain

HANDSHAKE SOLUTIONS

Asynchronous Circuits Adopted
Lessons Learned

Ad Peeters, May 15, 2007

Asynchronous circuits
Introduction

- It's safe to say 'asynchronous' has always had a relatively bad reputation
  - 'asynchronous' only specifies that it is *not* synchronous
  - This is often interpreted as undisciplined

- Yet, many groups are looking into disciplined approaches, for various reasons
  - <u>Universities:</u> asynchronous offers an infinite exploration space for research and publications
  - <u>Companies:</u> although they would typically want to stay away from asynchronous, this domain may provide solutions for energy consumption, variability and performance
  - Asynchronous even has its own conference! http://conferences.computer.org/async2007/

Asynchronous circuits
Status today

- Dozens of chip designs are mostly asynchronous
- More than 250 million of ICs on the market
- More than 80% of the world's smart (electronic, biometric) passports are based on asynchronous circuits
- Design point tools and complete design flows are now available
- Flows are connected to standard EDA flows
- Design-for-Test can be done using scan test
- Circuit overhead has been addressed/eliminated
- Several 'asynchronous' companies and start-ups

Asynchronous circuits Companies

- Theseus Logic
  - www.theseus.com
  - NCL (null convention logic) IP and services
- Fulcrum Microsystems
  - www.fulcrummicro.com
  - High-performance interconnect
- Handshake solutions
  - www.handshakesolutions.com
  - Tools, IP and services
- Silistix
  - www.silistix.com
  - On-chip interconnect
- Achronix Semiconductor
  - www.achronix.com
  - High-speed FPGAs

Asynchronous circuits
Acceptance issues

- Support and adoption by recognized companies
  - ARM, Boeing, Philips, NXP
- Recognition by technical media
  - EE Times, Microprocessor report
- Traction is visible in the market
  - Electronic passports
- Design tools
  - Design flows are connected to standard EDA flows
  - No special cells needed – standard-cell libraries supported
  - Design-for-Test can be done using scan test
  - Circuit overhead has been addressed/eliminated

HANDSHAKE SOLUTIONS

Asynchronous circuits adopted
1. Potential has been demonstrated

- Technology potential has always looked promising
  - Low power
  - High speed
  - Self adopting
  - Robust
  - No emission
  - Secure
- Yet these advantages cannot always be combined at will

- Proof points have meanwhile been created for some combinations

Technology benefits
Low current peaks and total current

ARM996HS consumes 2.8x less power than an ARM968E-S
and reduces current peaks by a factor 2.4

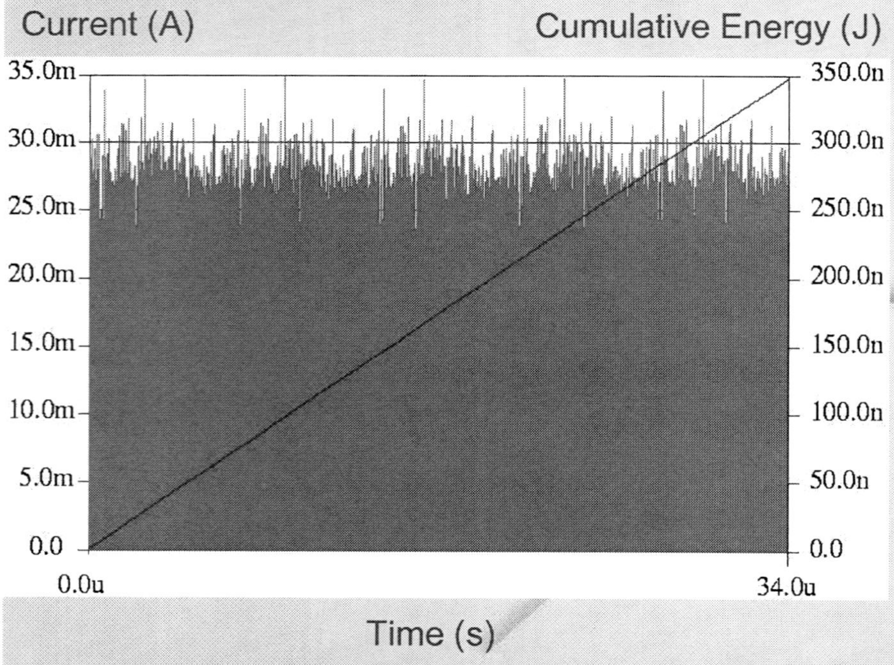

Current (A)   Cumulative Energy (J)

Time (s)

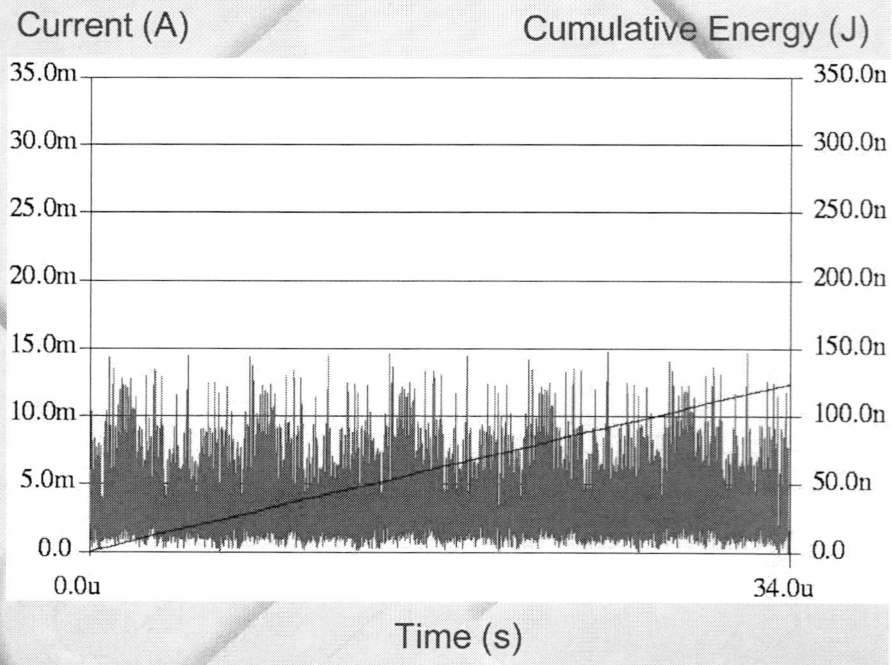

Handshake ARM996HS

Current (A)   Cumulative Energy (J)

Time (s)

Clock-gated ARM968E-S

Technology benefits
Current peak details

ARM996HS draws a relatively constant current, whereas the ARM968E-S current swings between 0 and 25-35 mA

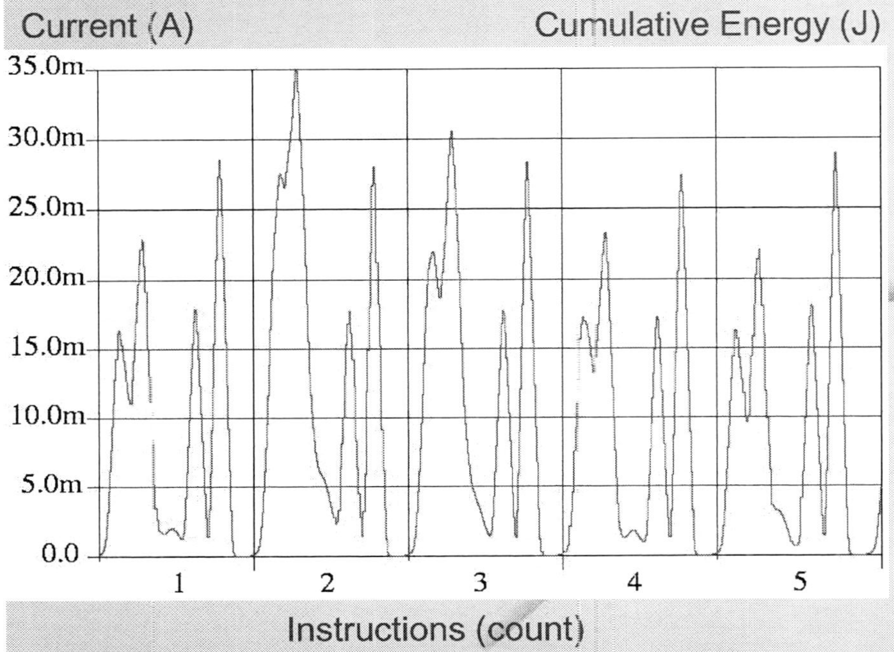

Clock-gated ARM968E-S

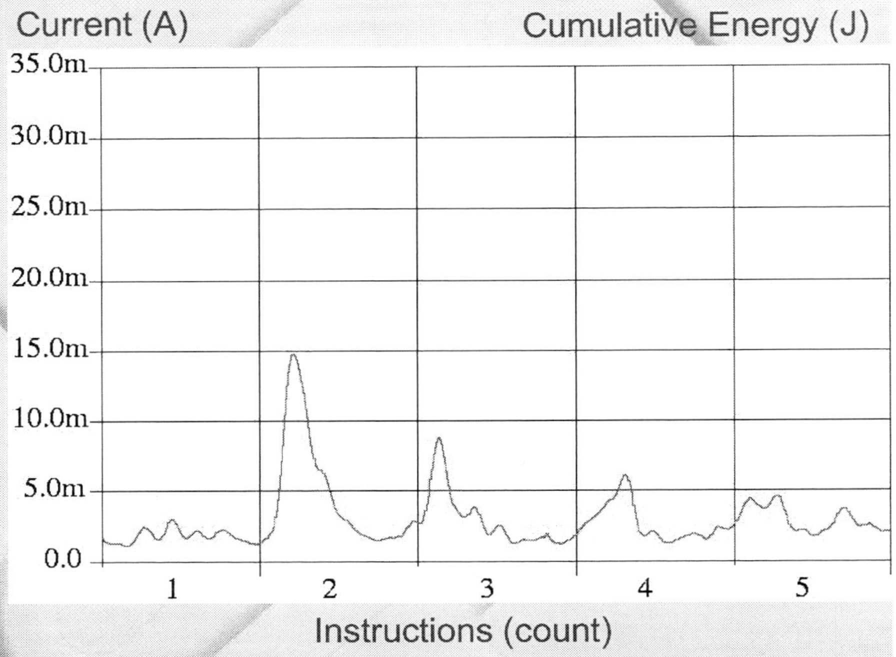

Handshake ARM996HS

Technology benefits
Current peak details – quiet region

ARM996HS draws a relatively constant current, whereas the ARM968E-S current swings between 0 and 15-25 mA

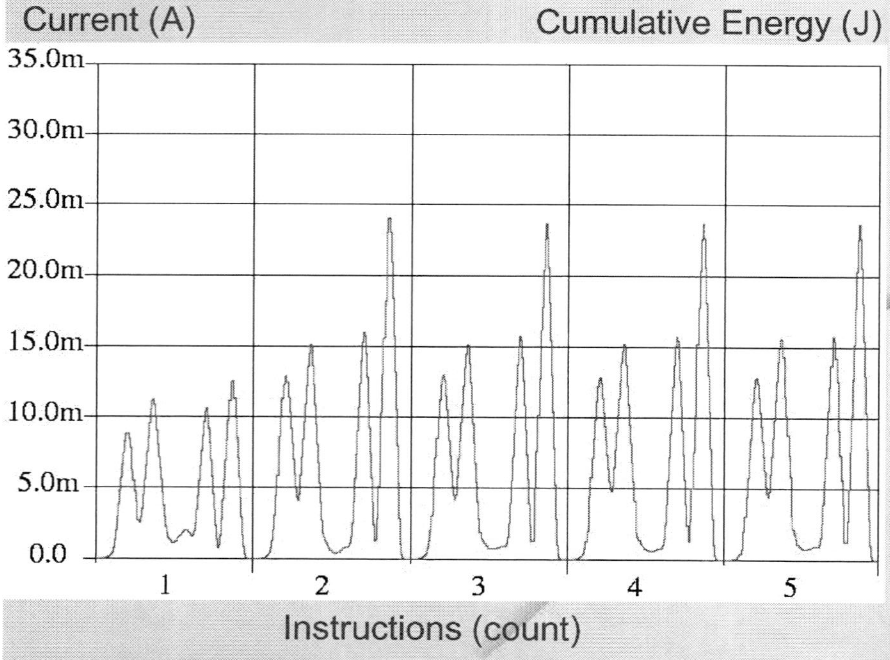

Clock-gated ARM968E-S

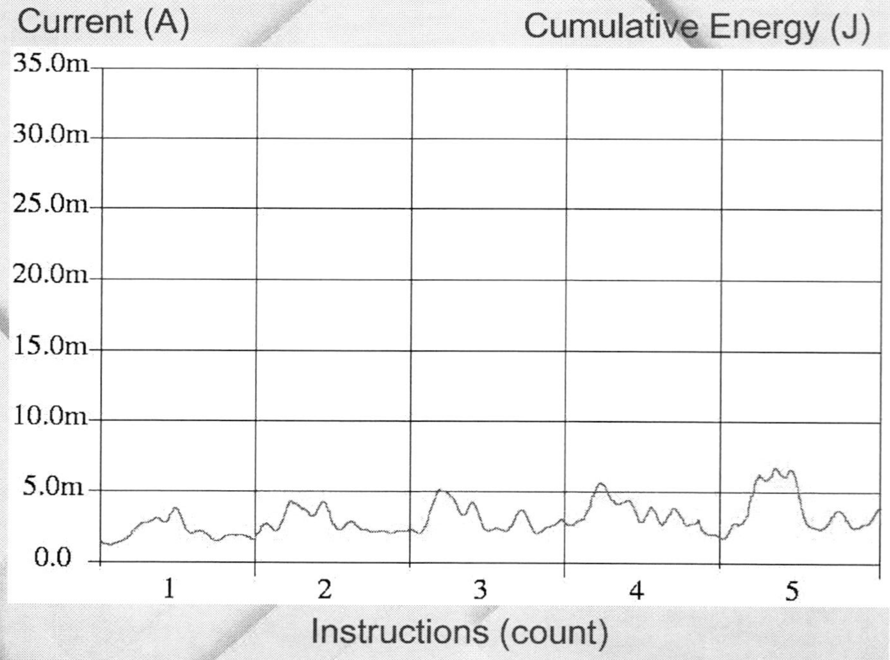

Handshake ARM996HS

HANDSHAKE
SOLUTIONS

Technology benefits
Current peak histogram

ARM996HS current typically between in 1 to 5 mA range, whereas ARM968E-S has significant current up to 20 mA

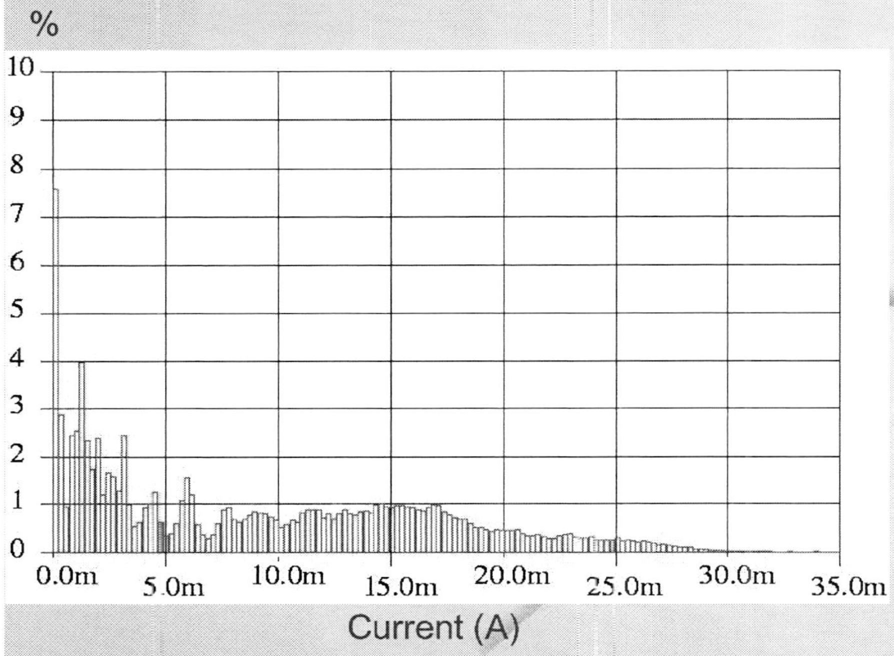

%

Current (A)

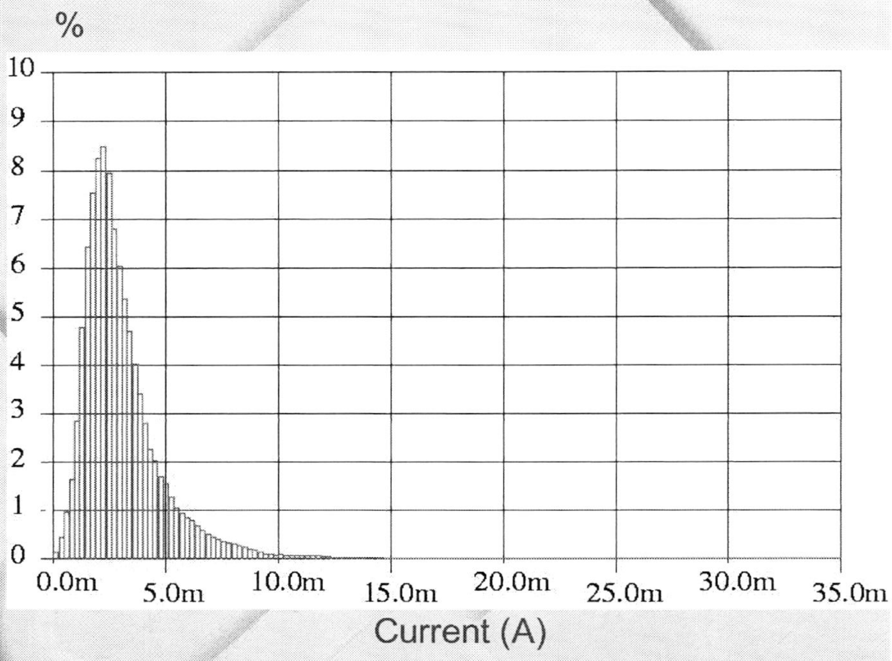

Handshake ARM996HS

%

Current (A)

Clock-gated ARM968E-S

Technology benefits
Low electromagnetic emission

ARM996HS offers low EME across the whole radio spectrum

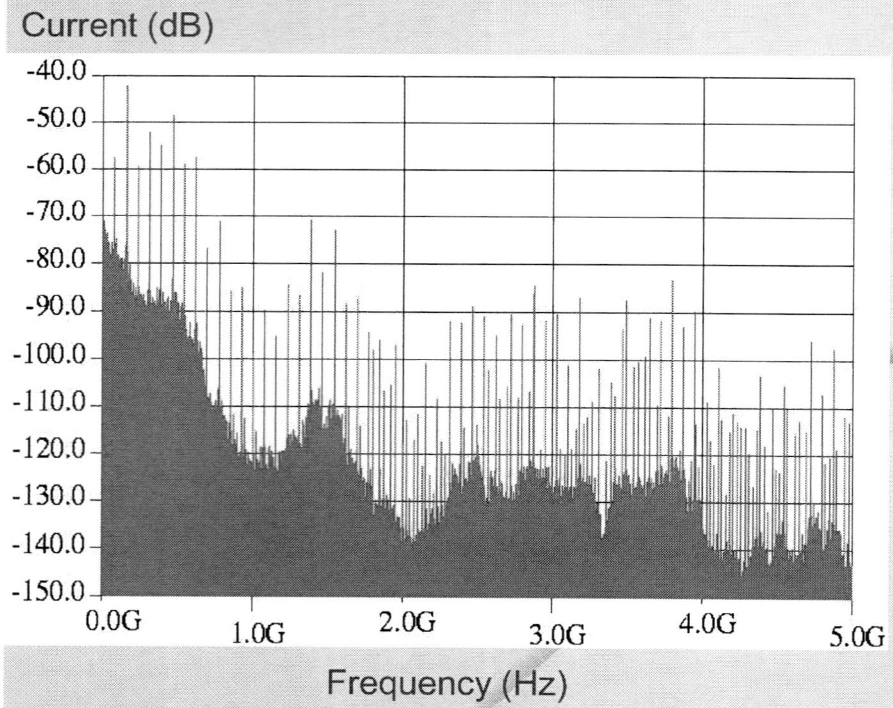

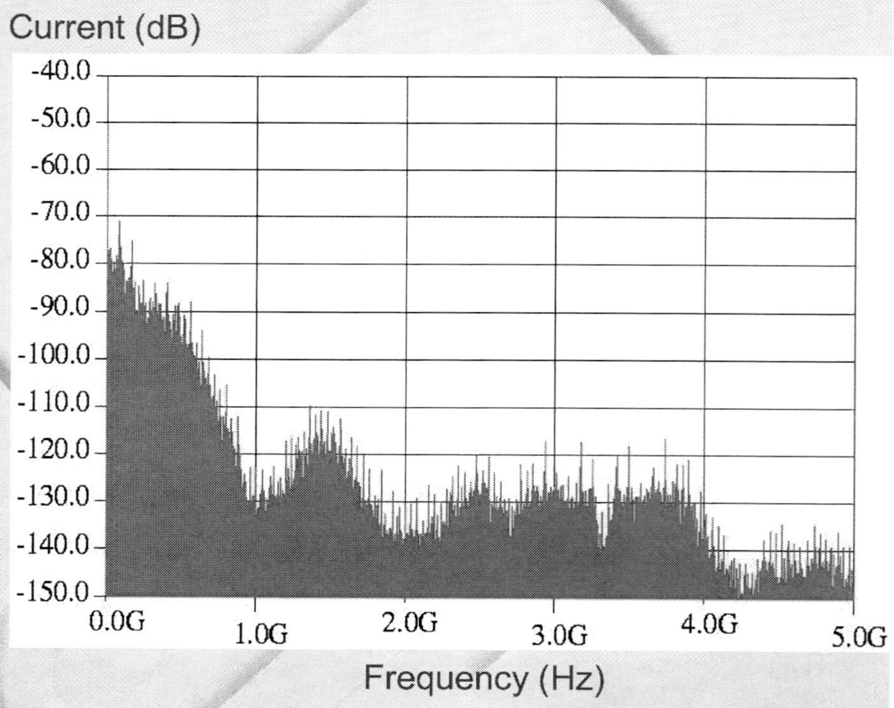

Handshake ARM996HS

Clock-gated ARM968E-S

Technology benefits
Low EME – FM radio band

ARM996HS guarantees no interference in FM radio band

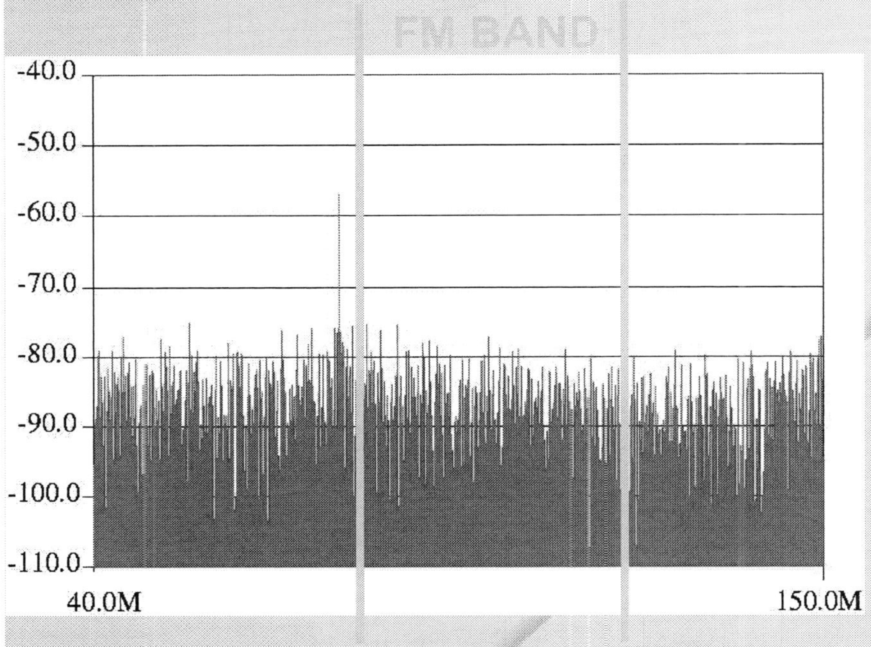

Clock-gated ARM968E-S at 77 MHz

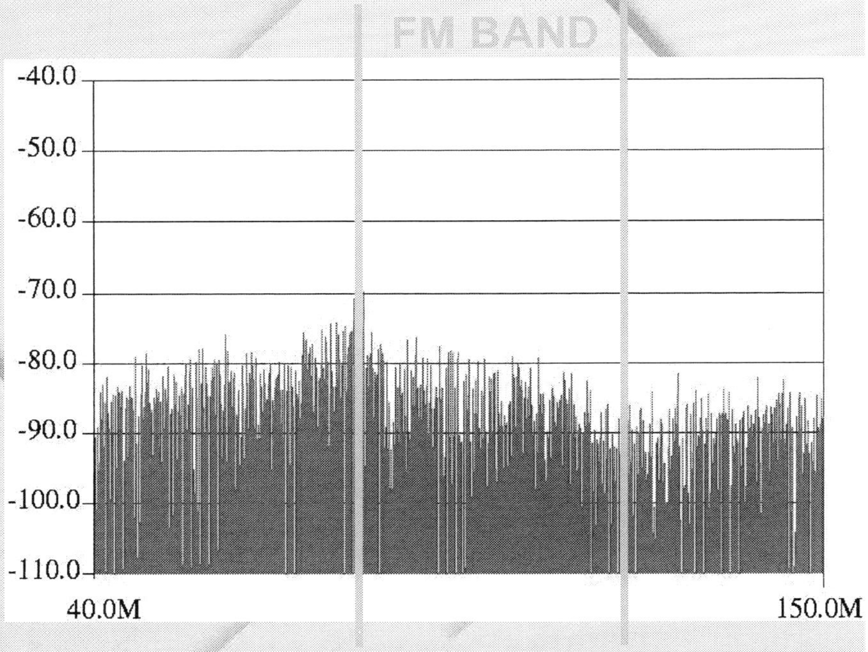

Handshake ARM996HS

Technology benefits
Low EME – FM radio band

ARM996HS guarantees no interference in FM radio band

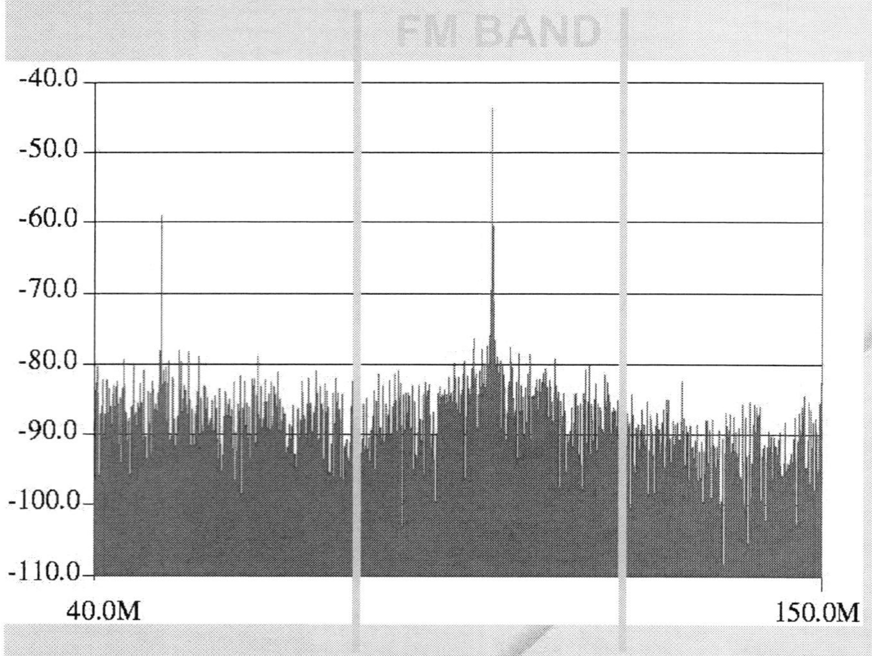

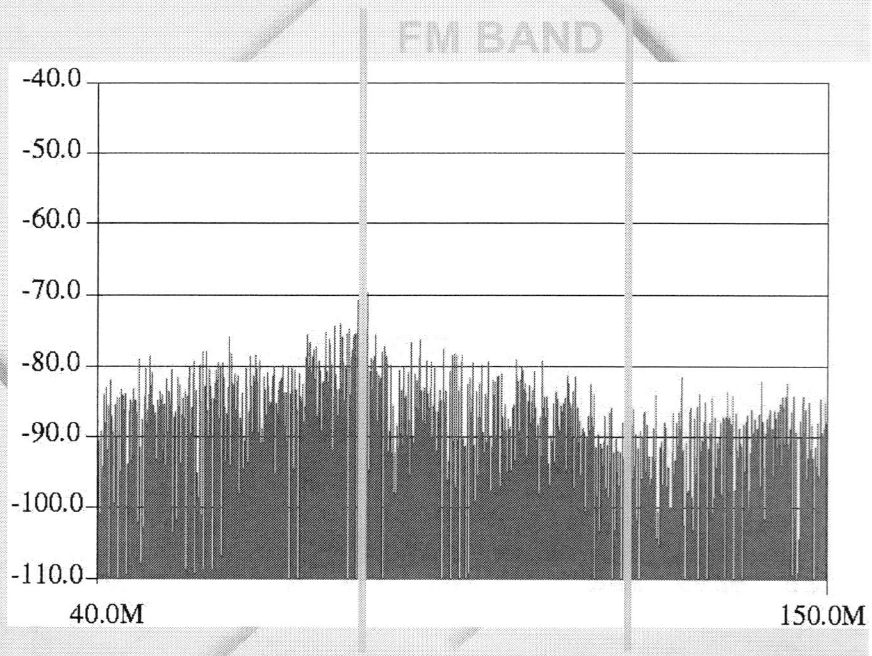

Handshake ARM996HS

Clock-gated ARM968E-S at 50 MHz

HANDSHAKE
SOLUTIONS

Technology benefits
Low EME – cellular band

ARM996HS reduces EME peaks by up to 25 dB in the
cellular band (800 MHz – 2.2 GHz)

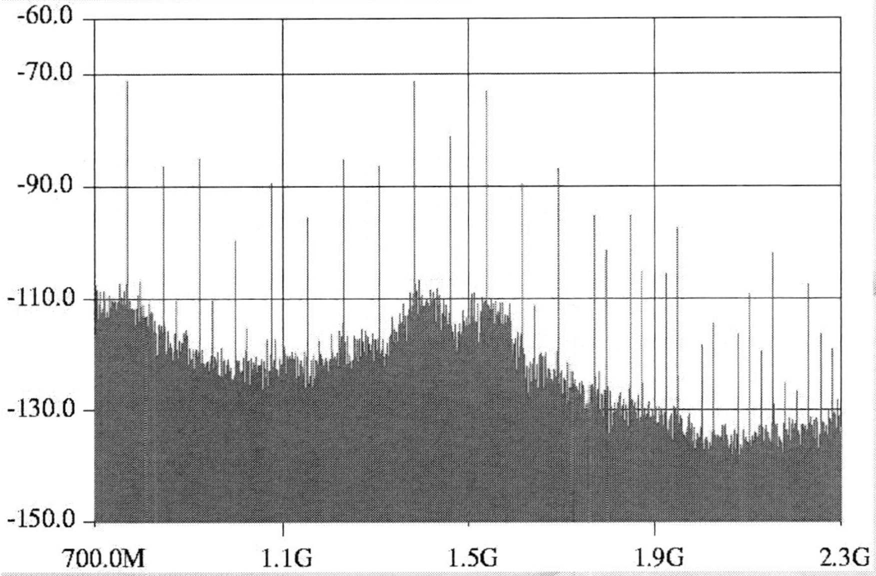

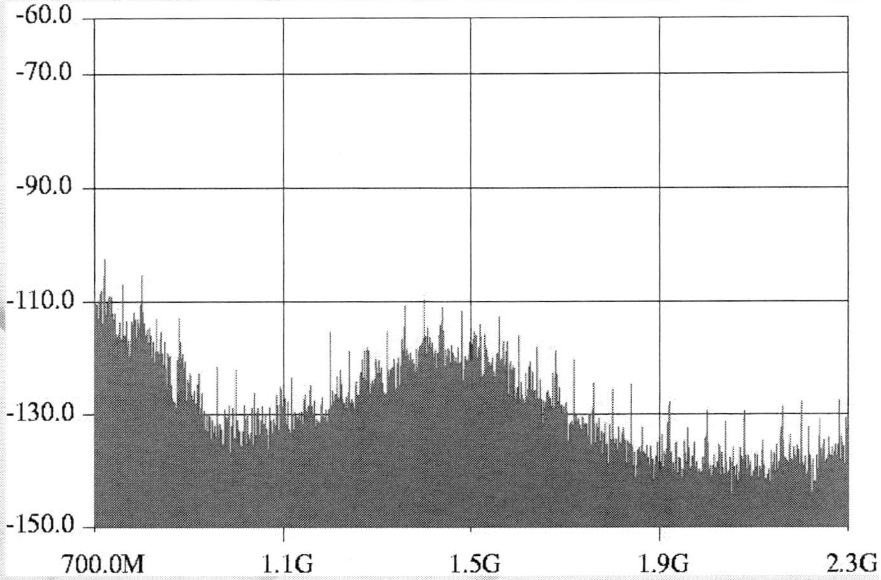

Handshake ARM996HS

Clock-gated ARM968E-S

Technology benefits
Low EME – wireless band

ARM996HS reduces EME peaks by up to 30 dB in the wireless band (2.2 GHz – 2.6 GHz)

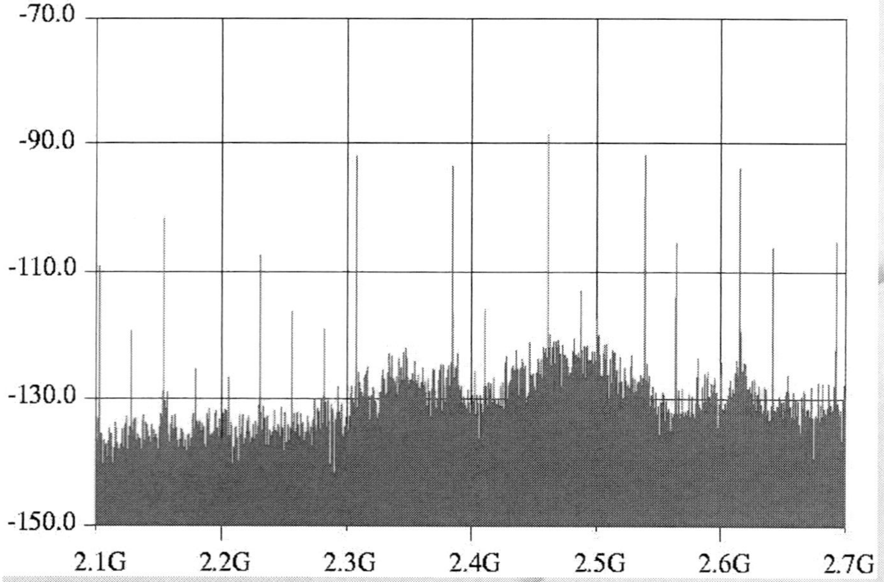

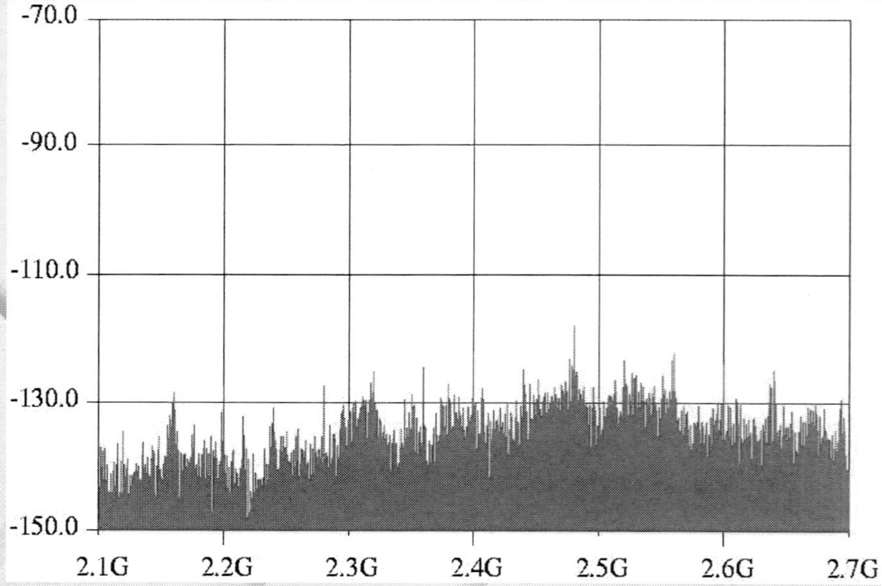

Handshake ARM996HS

Clock-gated ARM968E-S

Asynchronous circuits adopted
2. Products on the market

- Benefits are nice to have,
  - ... but these need to be translated into product advantages
  - ... and brought to market

- Several dozens of ICs done
- More than 250 million ICs sold

  - Examples follow

Products on market
Communication – motivated by emission

Multi-standard pager

Game controller

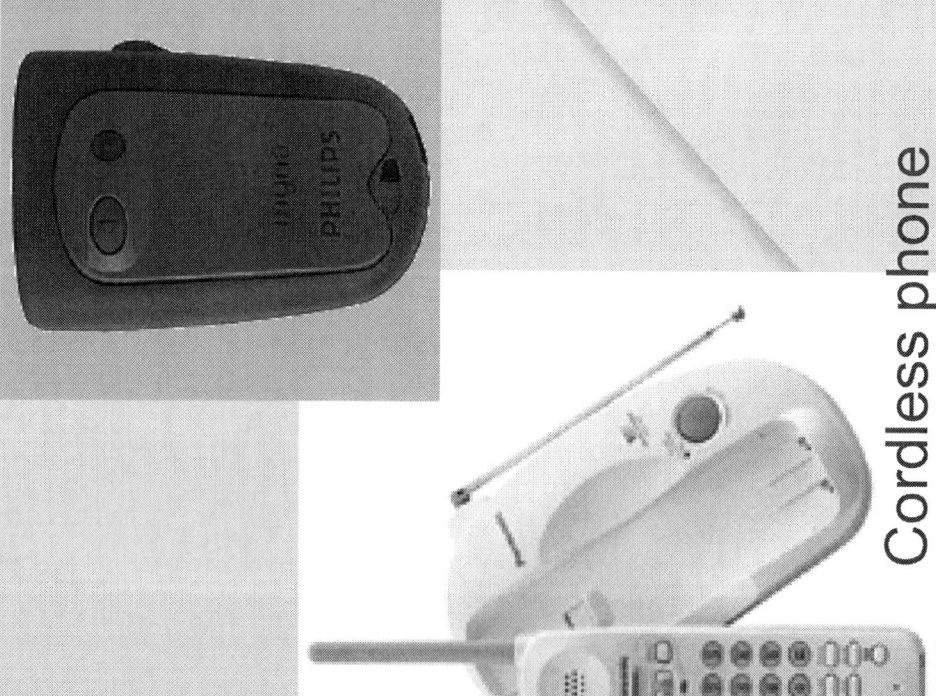

Cordless phone

Products on market
Smartcards – motivated by energy efficiency

Asynchronous (handshake) circuits power more than 80% of the world's smart passports

Nokia 6131 NFC phone

Products on market
Automotive – motivated by robustness

MEMS
- Tire pressure sensors

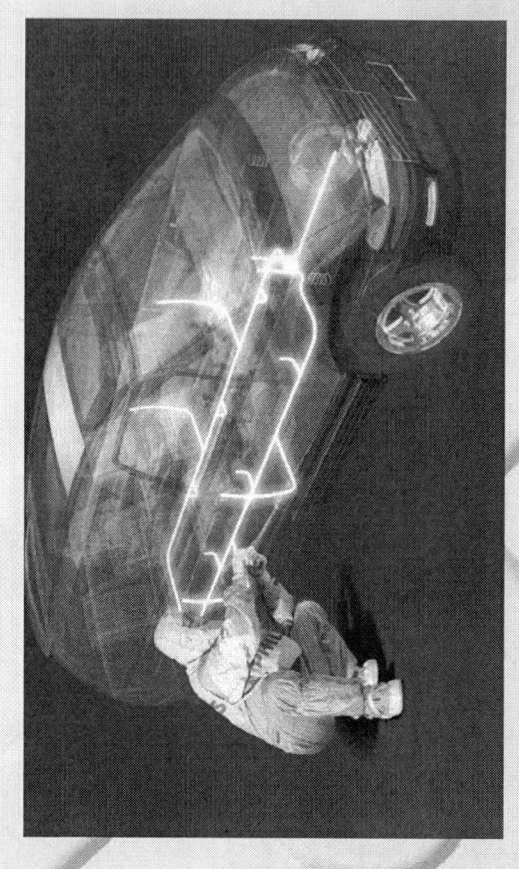

In vehicle networking:
- CAN and LIN
- Transceivers and controllers

HANDSHAKE SOLUTIONS

Asynchronous circuits adopted
3. Design tools available

- Timeless Design Environment is a frontend to 3rd-party EDA flows
- TiDE is *complementary to and compatible with third-party EDA* tools
- High-level design entry (Haste)
- Standard-cell hand-over
- Scan-test-based Design-for-Test
- FPGA prototyping through synchronous preview of design
- Integrated support for placement and routing, logic optimization and timing sign-off

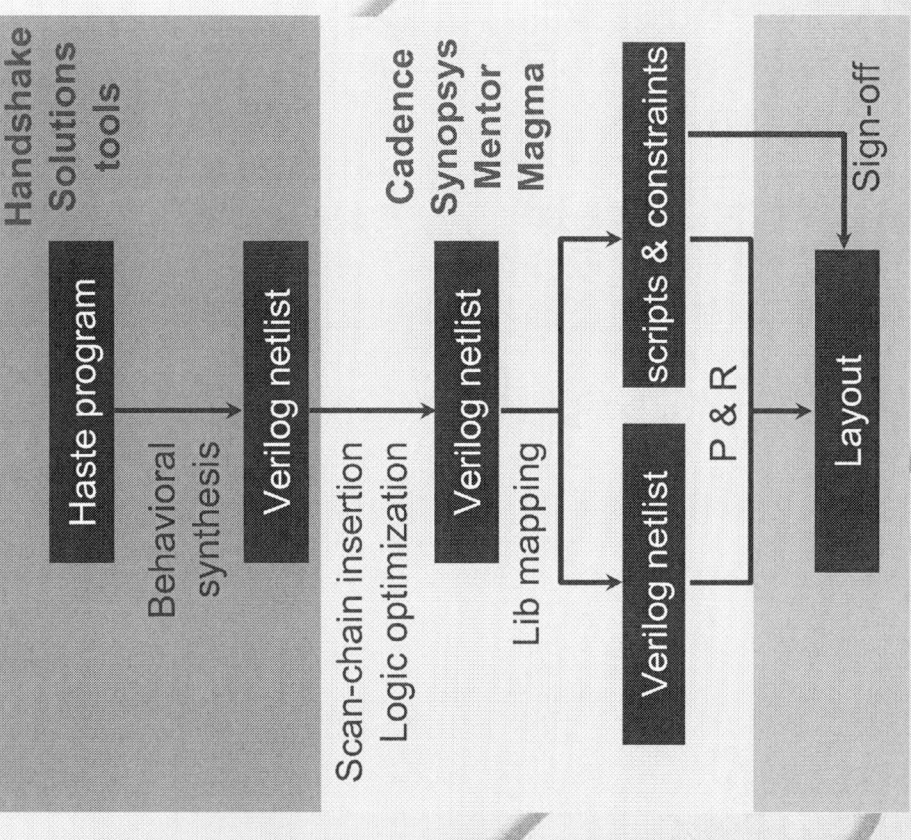

Design tools available
Mapping on standard cells

<u>Muller's C-element:</u>
Set-reset gate with symmetric
set/reset function

$$a \wedge b \;\rightarrow\; z:=1$$
$$\neg a \wedge \neg b \;\rightarrow\; z:=0$$

Can be implemented by majority gate with feedback

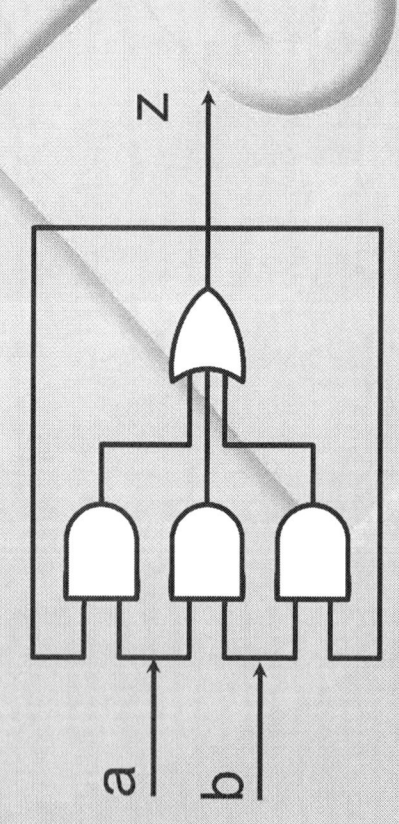

$$z=a*b+z*(a+b)$$

Design tools available
Design-for-Test based on full scan

- 'Synchronous' test quality for asynchronous circuits
  - DfT based on (synchronous) full scan
  - Full coverage against stuck-at-faults, bridging faults, IDDQ

- Circuit is transformed into a synchronous scanable design
  1. Test multiplexer and multiplexed test clock for datapath registers
  2. Combinational loop breaking in control circuit using multiplexers
  3. Reuse datapath scan to control and observe control circuit

- Test pattern generation using standard tools
  - Based on remodeled views of asynchronous circuit (after layout)
  - Remodeling is minimized, needed to work around ATPG tool limitations
  - Standard ATPG tools: Tetramax, Fastscan, EncounterTest, Amsal

- Proven solution e.g. in automotive domain
  - ARM996HS at 99.3% stuck-at coverage

Asynchronous circuits adopted
4. Rewards and visibility

- EE Times, ARM996HS #2 in top stories 2006

- MPR innovation award for ARM and Handshake Solutions

- Achronix Completes $25.4 Million Series A Preferred Stock Financing

- Silistix honored in EDN's "Hot 100 Products of 2006"

- Fulcrum Microsystems and Applied Micro Circuits Corporation demonstrate first low-latency 10-Gigabit SFP+ switch

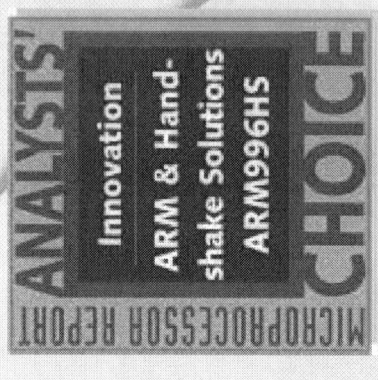

- Lowest latency (200ns)
- Most power efficient (<100mW/Gbps)
- Most integrated (single chip)

Per-port configurable:
10G (XAUI, CX4)
10/100/1000, 2.5G (SGMII, CX)

FocalPoint

24 Ports

HANDSHAKE SOLUTIONS

Asynchronous circuits adopted
5. Design teams converted

- In contrast to common believe, most engineers like change and innovation
  - Even learning a new design language is not a handicap per se
- Teams learn, even when they don't convert to asynchronous
- A 'considered' way of designing and engineering is stimulated
  - Awareness of design alternatives that influence power, energy efficiency, flexibility

Asynchronous circuits adopted Conclusion

- Asynchronous circuit technology is disruptive
  - Adoption will go via continued evolution, not a revolution
  - Adopt to standard wherever possible helps

- The march forward is ongoing
  - Many parties are interested
  - Many are even investigating
  - Several are doing design starts
  - >250 Million ICs on the market today

Asynchronous circuits accepted
Where are we now on the hype curve?

HANDSHAKE SOLUTIONS

Visibility

Key: Time to Plateau
- ○ Less than two years
- ◐ Two to five years
- ● Five to 10 years
- △ More than 10 years

Carbon Nanotube Circuits △
Phase Change Memory ●
Network on Chip ●
Plastic Transistors ●
GaN Devices ●
3D Wafer Stacking ●

Optical Silicon ●
Polymer Memory △
Molecular Transistors △
Quantum Computing △
DNA Logic △

'Inkjet' Manufacturing ●
Silicon RF power amplifiers ●
HW Reconfigurable Devices ●
Digital paper ●
EUV Lithography ●
Micro Fuel Cells ●
Multi-gate Transistors ○

Embedded FPGA Cores ●
MRAM ●

LCoS ●
Analog IP ●
Strained Silicon ●
Asynchronous Logic ●
Organic Light Emitting Devices ●
ESL Tools and Methodologies ●

FRAM ○
Immersion Lithography ○

RF ○
CMOS ○

Trigger | Peak of Inflated Expectations | Trough of Disillusionment | Slope of Enlightenment | Plateau of Productivity

Source : Gartner 2006 Industry Review FSA / HS interpret.

HANDSHAKE SOLUTIONS

Thank you

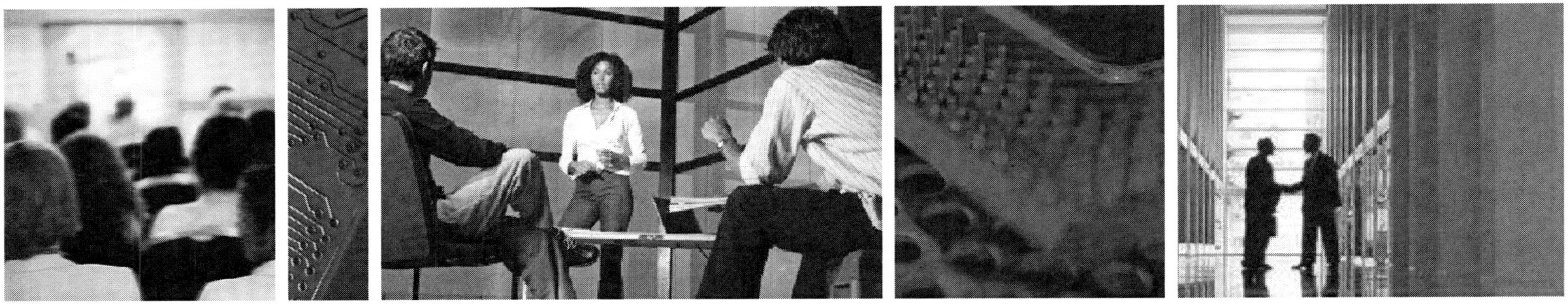

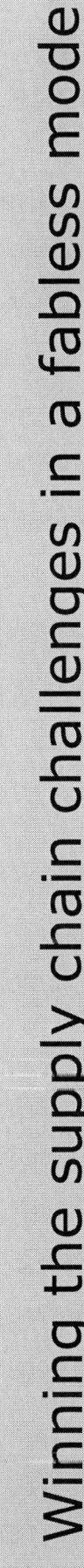

Winning the supply chain challenges in a fabless model

Ron Torten
CEO, Nemerix
May 15, 2007

Agenda

FSA

- Nemerix Introduction
- Why is fabless different
- Key industry trends impacting semiconductor companies
- What does this mean
- Winning the challenge
- Summary

Nemerix Overview

- Privately held fabless semiconductor company
- HQ in Manno, Switzerland
- Leader in low power GPS/AGPS solutions for handheld application
- Founded in 2002
- Sales to date >$20M and >6M units

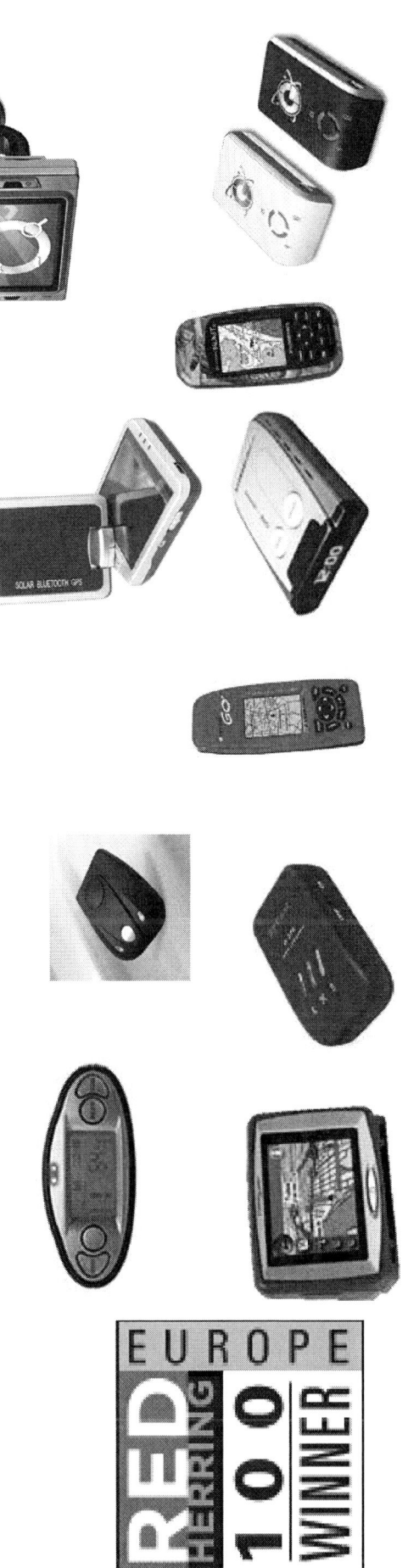

Why is Fabless Different

- Limited impact on process
  - Business case has to be justified by external measures
  - Must work well by design
  - DFM vs. Processing for design
- Dependency on external resources with their own priority
  - Limited visibility to what other priorities exist
- Distance
  - Manufacturing is usually in a different time zone
- Competition with external factors for capacity and pricing
  - You will only get what the supplier believes you should get
- Supply chain and supplier management as a core corporate competence is required.

What is changing

- Consumer market became the largest Semiconductor market

- Cost pressures driving integration and increasing IP complexity

- Technology migration increasing complexity and design cost

- Semiconductor industry less attractive for VC investment

- Success of Fabless business model attracting IDM to transition

- A few "mega" fabs dominate the market

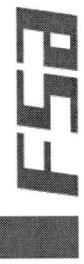

Change in End Market

- **Consumer market dominating Semiconductor demand**
- **TTM implication of profitability growing dramatically**
- **ASP declines drives requirement for high volumes**

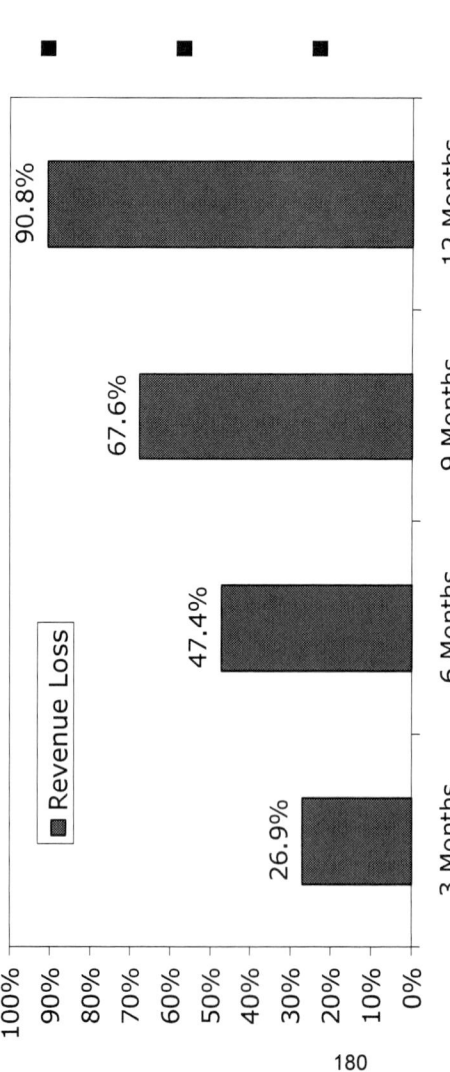

Revenue Loss for Being Late to Market

- Revenue Loss

26.9% — 3 Months
47.4% — 6 Months
67.6% — 9 Months
90.8% — 12 Months

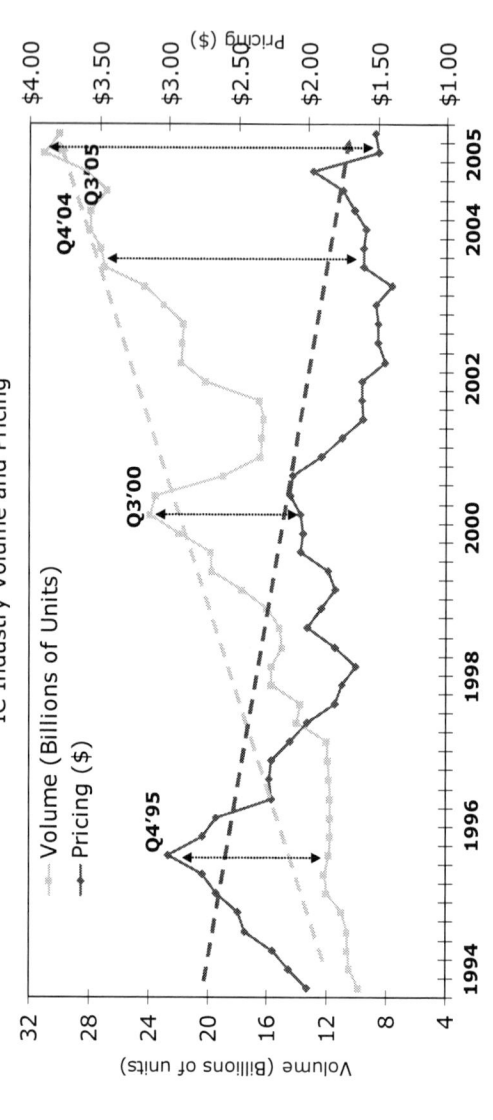

IC Industry Volume and Pricing

— Volume (Billions of Units)
— Pricing ($)

Q4'95 Q3'00 Q4'04 Q3'05

Source: IBS, 2007

Change in Technology

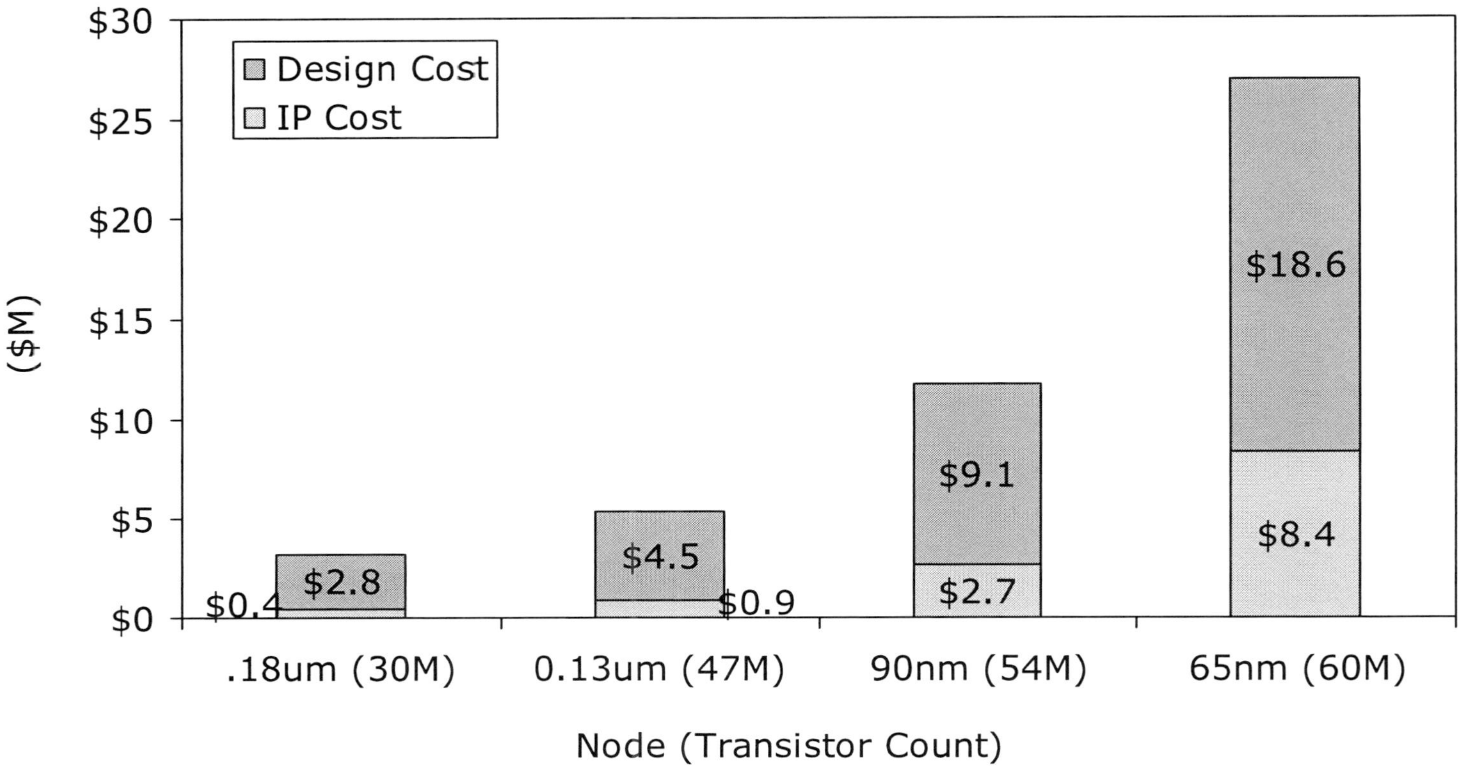

Design Cost · **IP Cost**

($M)

Node (Transistor Count)	Design Cost	IP Cost
.18um (30M)	$2.8	$0.4
0.13um (47M)	$4.5	$0.9
90nm (54M)	$9.1	$2.7
65nm (60M)	$18.6	$8.4

As nodes advance both design and IP costs increase exponentially

Source: IBS, 2007

Change in Funding Model

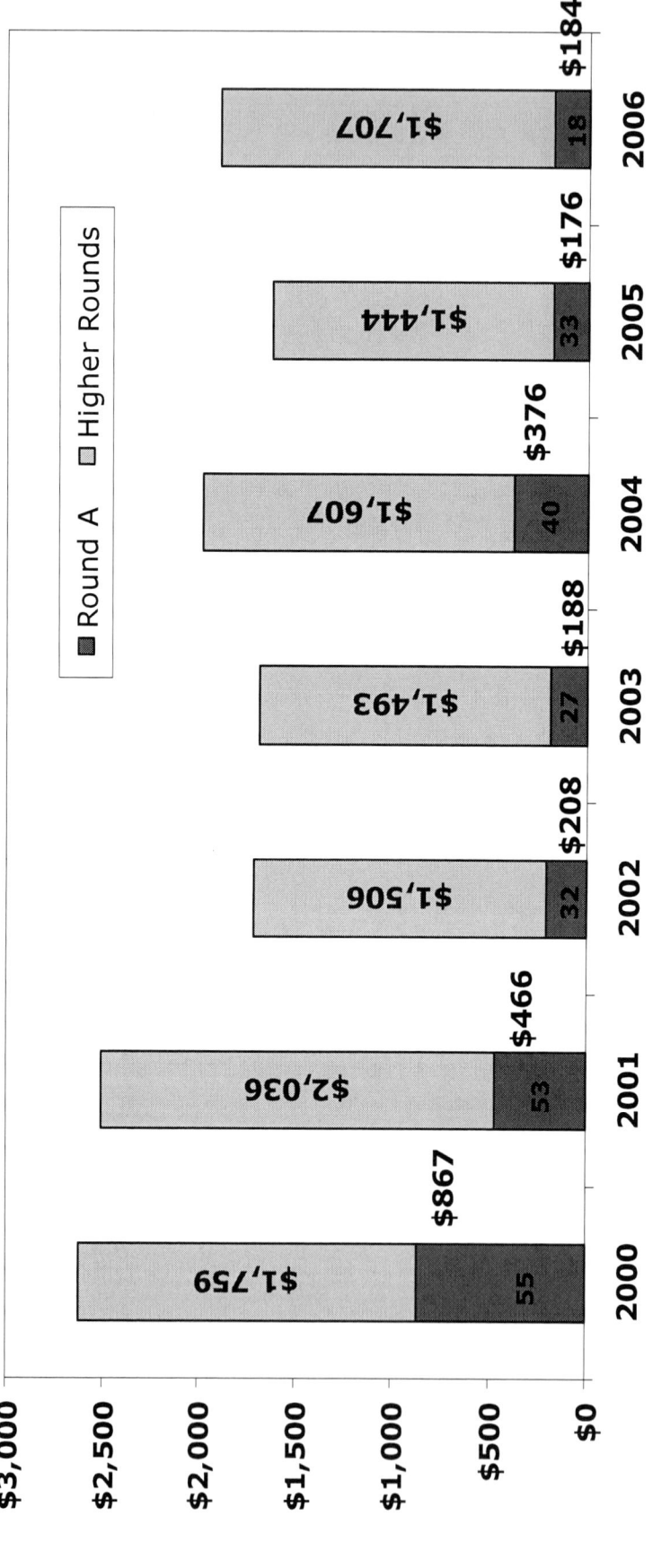

Year	Round A	Higher Rounds	Total
2000	55	$1,759	$867
2001	53	$2,036	$466
2002	32	$1,506	$208
2003	27	$1,493	$188
2004	40	$1,607	$376
2005	33	$1,444	$176
2006	18	$1,707	$184

Round A deals have declined nearly 80% between 2000-2006

Return to both founders and VC are more difficult

Change in Profitability

- Fabless companies have a more profitable cost structure than IDMs

- Fabless companies have a better average EBITA margin than the IDMs

- R&D and SG&A are similar as a percentage of revenue

- Significant portion of the foundries EBITA contribution is due to TSMC alone

Aggregate industry cost structure, 2005
% of sales

	Foundry	IDM	Fabless
COGS	32.3	48.5	53.0
Depreciation	38.4	16.2	2.0
		15.2	16.0
R&D	7.1	11.1	10.0
SG&A	5.1		
EBITA	17.2	9.1	19.0

Source: *McKinsey CPC Semiconductor Database*

Fabless Model Outperforms IDMs consistently
Gap grows with increase mfg. costs at advanced technologies

Change in growth rate

FSA

Fabless Industry

Semi Industry

($M)

$60,000
$50,000
$40,000
$30,000
$20,000
$10,000
$0

($M)

$300,000
$250,000
$200,000
$150,000
$100,000
$50,000
$0

— Semi Industry — Fabless Industry

Fabless CAGR = 26%
Semiconductor CAGR = 8%

$49.6B
$247.7B
$204.4B
$17.0B

1987 1988 1989 1990 1991 1992 1993 1994 1995 1996 1997 1998 1999 2000 2001 2002 2003 2004 2005 2006

Fabless growing 3X faster that Semi overall

Source: FSA

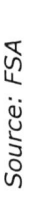

Change in size limitation

	Company (Highlighted companies–Non U.S.)	Stock	Ticker	CY 2006 Revenue ($000)	CY 2005 Revenue ($000)
1	QUALCOMM (QCT Division)	NASDAQ	QCOM	$4,331,000	$3,457,000
2	Broadcom	NASDAQ	BRCM	$3,667,818	$2,670,788
3	SanDisk Corporation	NASDAQ	SNDK	$3,257,525	$2,306,069
4	NVIDIA Corporation	NASDAQ	NVDA	$3,068,771	$2,375,687
5	Marvell Technology Group Ltd	NASDAQ	MRVL	$2,237,596	$1,670,266
6	LSI Logic	NYSE	LSI	$1,982,148	N/A**
7	Xilinx, Inc.	NASDAQ	XLNX	$1,871,604	$1,644,890
8	MediaTek Incorporation	TSEC	2454	$1,624,486	$1,410,986
9	Avago Technologies	Private	Private	$1,600,000*	$1,800,000
10	Altera	NASDAQ	ALTR	$1,285,535	$1,123,739
11	Conexant Systems	NASDAQ	CNXT	$985,615	$812,818
12	NovaTek	TSEC	3034	$964,314	$788,558
13	Himax Technologies	NASDAQ	HIMX	$744,518	$540,204
14	CSR	LSE	CSR.L	$704,700	$486,531
15	VIA Technologies, Inc.	TSEC	2388	$657,901	$580,698

*Avago's 2006 net revenue accounts for the 12 months ended Oct 31, 2006.
**LSI Logic not fabless until 2006.

Source: FSA

Foundry Market share 2006

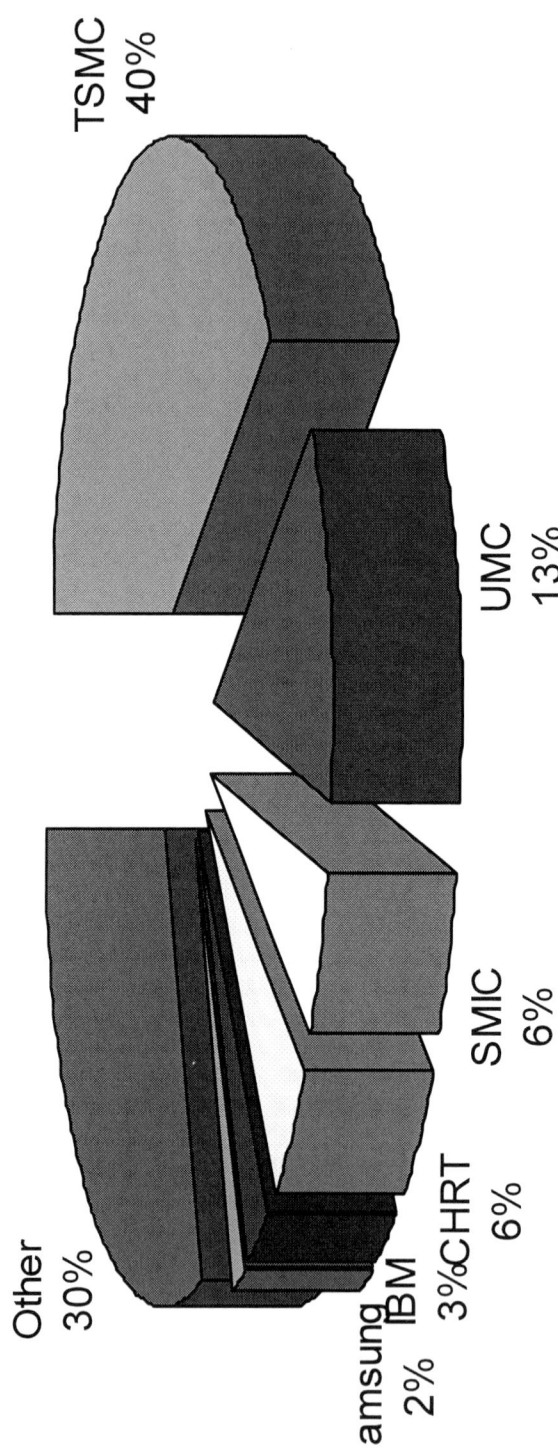

TSMC 40%

UMC 13%

SMIC 6%

CHRT 6%

IBM 3%

Samsung 2%

Other 30%

Change in IDM strategy

FSA

- Agere/LSI – Planned for fabless
- Avago/HP – Moved to fabless
- TI – Migration with advanced technology
- Freescale - Migration with advanced technology
- IBM – Enabled via cooperation with Samsung/CHRT
- Other have mixed model – IDM + Outsource

Investment threshold increases from $2B rev. in '00 (Agere/HP) for 90nm to higher rev. required for development of advanced technologies

What does this mean?

- Unless you are big ($2B+), influence on process

- With IDMs moving to fabless, mid-size companies will be smaller on a relative scale

- Pricing power is reduced

- Differentiation based on process or cost is challenged

- Allocation discussions, if required will be more difficult

- With the consolidation of business with a few foundries, they gain market intelligence

Winning the supply chain challenge

- Differentiated relationships will be based on strategic alignment

- Move from supplier relationship to partner

- Move from partnership of convenience to true Win/Win

- Continually think about how you, as a foundry customer you stand out in the crowd of customers
  - Over 1000 fabless companies WW

- Increase investment in process engineering will be required – if you can afford it
  - If not, focus on where you can differentiate

- Supplier relationship management must become a corporate focus
  - Goes well beyond pricing and capacity

Examples

FSA

- Shared IP
- Co-investment
- Technology development
- IDM capacity transition
  - New Markets
  - New Technology
- New markets

Summary

- Fabless market and supply chain is changing
- Differentiation in the eyes of the foundry will be critical and requires focus
- For many companies, this will be a difficult conceptual transition
- Those that will figure it out will have better tools to compete

DFY, DFT and DFM: A Pragmatic Approach to Built-in Quality

Graham Curren
CEO

About Sondrel

- IC implementation design service provider with key services in:

  - Verification, STA, Synthesis, Test, P&R
  - Analogue and full custom layout
  - Library and IP sourcing and qualification
  - From netlist to production management
  - Design optimisation

- Established 2002; Annual growth 45%

- > 20 designs per year, from 65nm to 180nm

- Located throughout Europe and Israel

Design For Rol

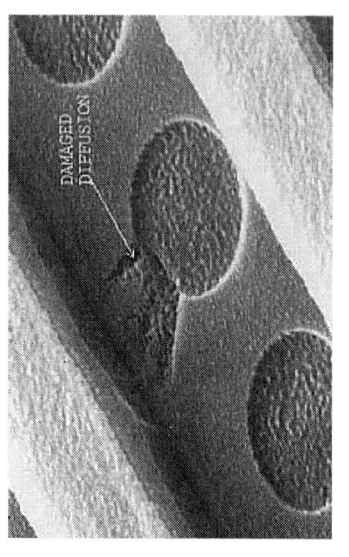

DAMAGED DIFFUSION

Contact too close to poly gate leaves no space to vent heat

1. Can the results be measured?

2. Have key tasks been prioritised?
   - Smallest die area
   - Test strategy
   - Timing closure and design quality
   - Time to market or time to schedule?

3. Design in quality, or throw away defective parts?
   - Trade off NRE and unit cost

What Affects Yield?

- ## Design Quality
  - Synchronous design, redundancy

- ## Verification
  - Formal, functional, STA

- ## IP
  - Quality, performance, modelling

- ## Manufacture
  - Stability, drift, development

- ## Environment
  - Heat, voltage

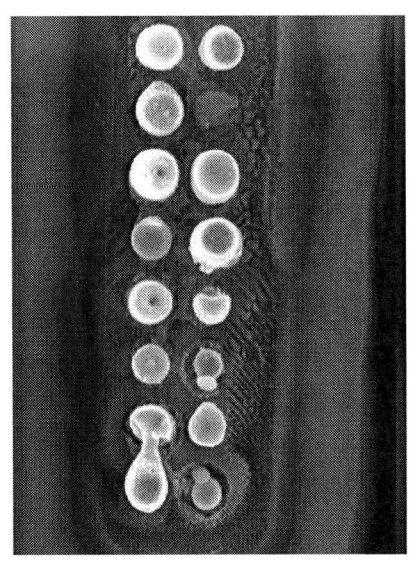

Burned out via chains
0.15um technology

Design For Test

- Chip quality improves faster due to better test coverage than better yield

- Testing will achieve improvements in chip quality at lower costs

- Get the most out of DFT

  - Use a test expert to define your test strategy up front:
    - Test modes
    - Fault models
    - Tester time (cost)
    - Scan compression
    - etc

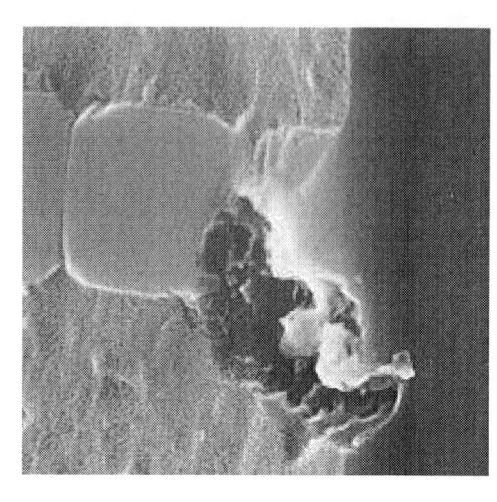

Gate oxide damage

Best Practice

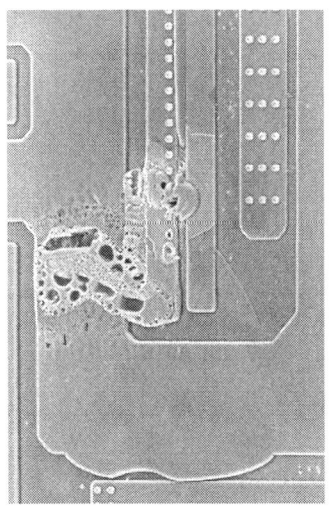

Multiple problems due to thick metal

- Address the fundamental, high-return issues first

  - Quality of design
  - Quality of IP and libraries
  - Quality of verification
  - Quality of test strategy
  - Quality of chip finishing
  - Time to market

- Understand the financial implications

- Work closely with all suppliers (IP, foundry, package, test etc)

- If it's not measurable or stable, is it worth doing?

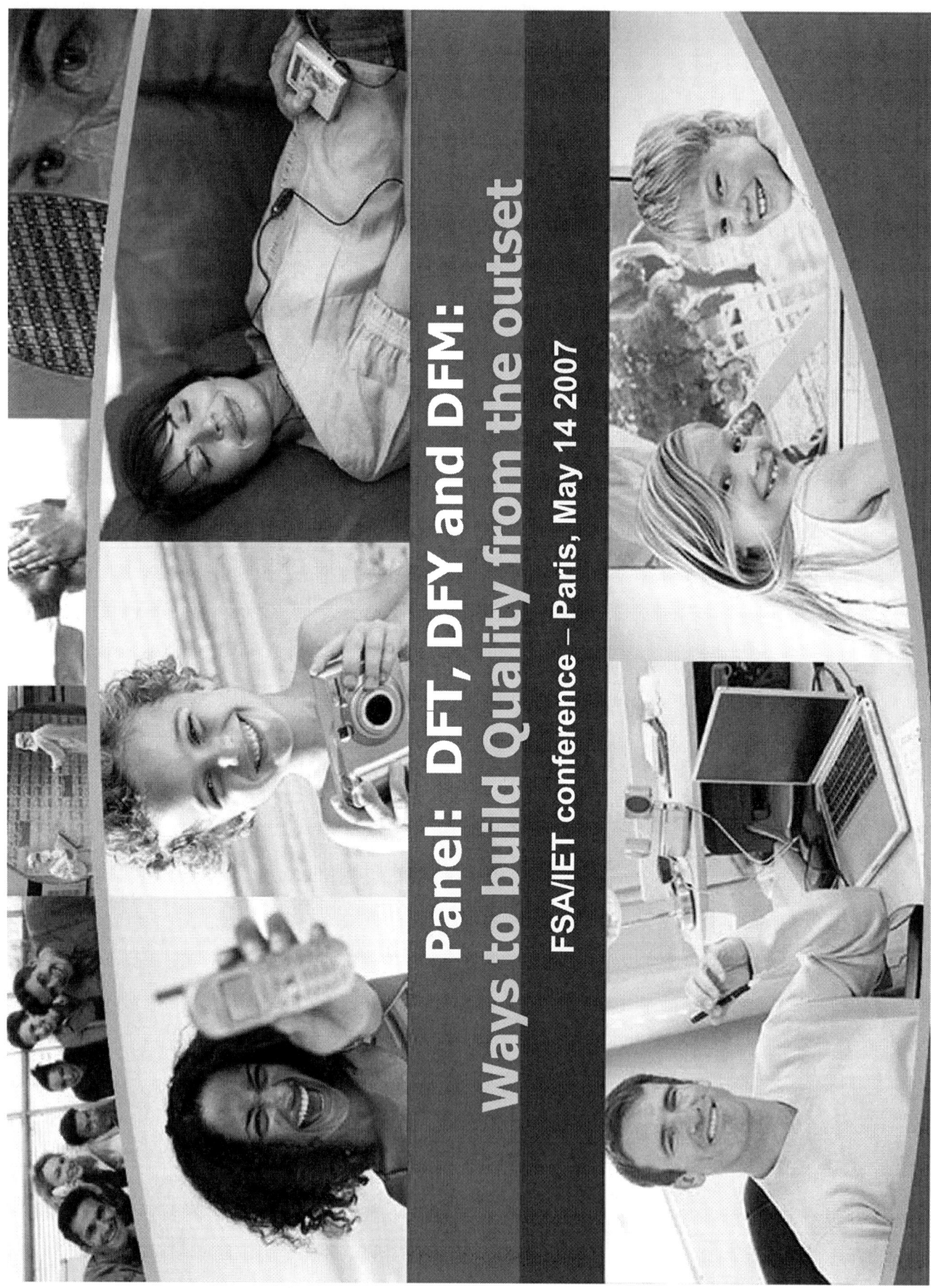

**Panel: DFT, DFY and DFM:
Ways to build Quality from the outset**

FSA/IET conference – Paris, May 14 2007

What does DFM cover?

Design

Technology

- Architecture
- Design
- Data Patterning
- Mask Patterning
- Wafer Patterning
- Silicon Process
- Test
- Package
- Reliability
- Validation

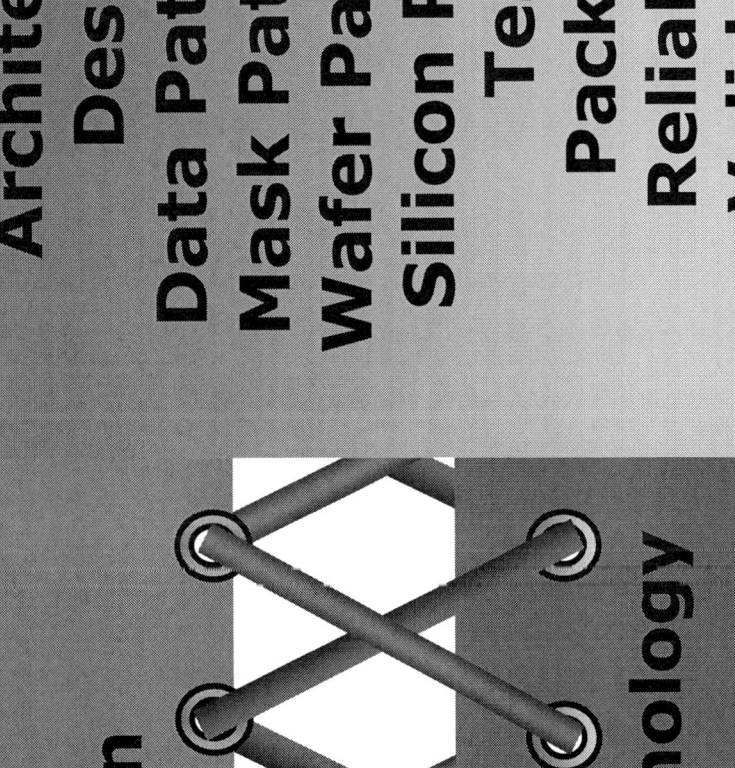

Based on recent Intel presentation

What is DFM?

(3) Mitigate
- Architecture
- Circuit
- Layout
- Design Tools & Methods
- Design Rules

(1) Define
- Models
- Sensitive circuits
- Data Patterning

(2) Control
- Control & Analysis Systems
- Process Recipes
- Equipment
- Materials

Based on recent Intel presentation

Traditional Cooperation Between Design and Manufacturing

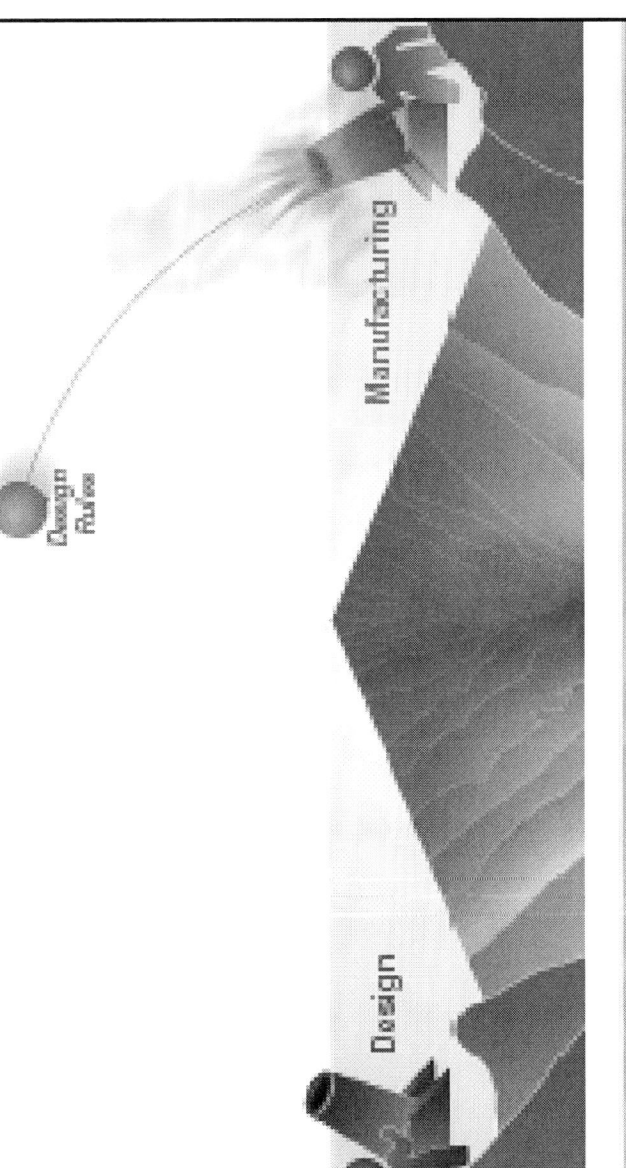

From a recent Mentor presentation

Traditional Cooperation Between Design and Manufacturing

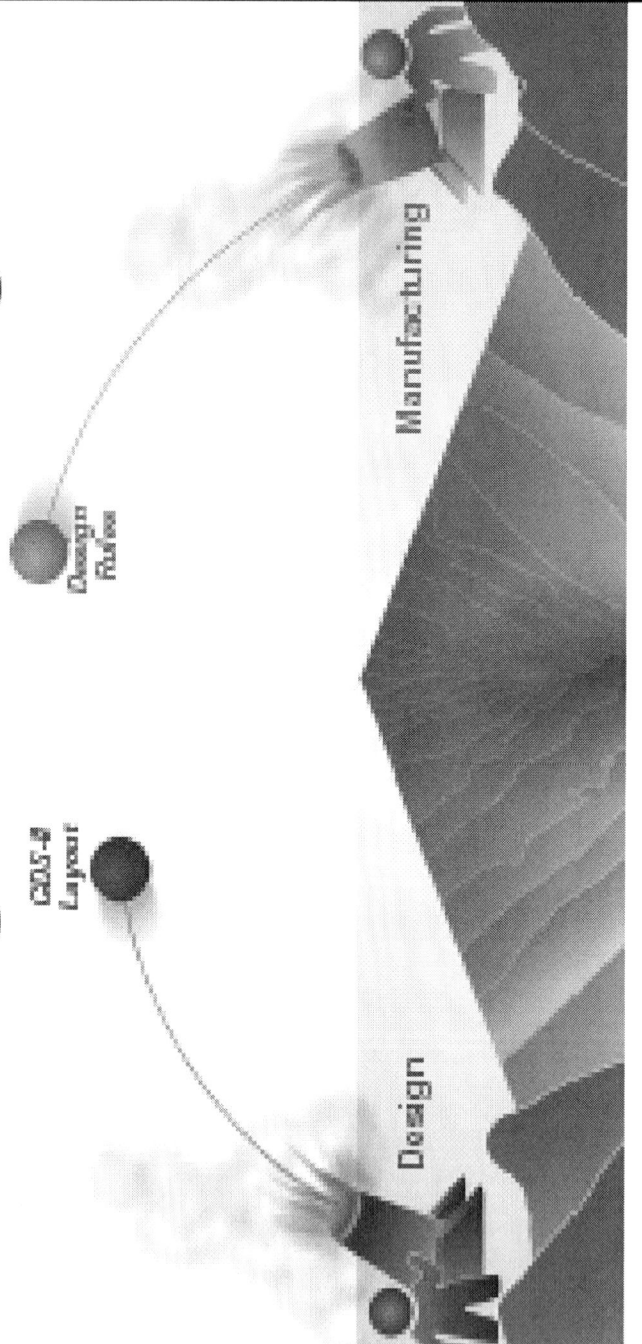

From a recent Mentor presentation

Design Foundry

Litho Friendly Design

From a recent Mentor presentation

borrowing from Mike Splinter's vocabulary*...

*"Is technology going out of fashion", 1/10/06, SEMI ISS, CA

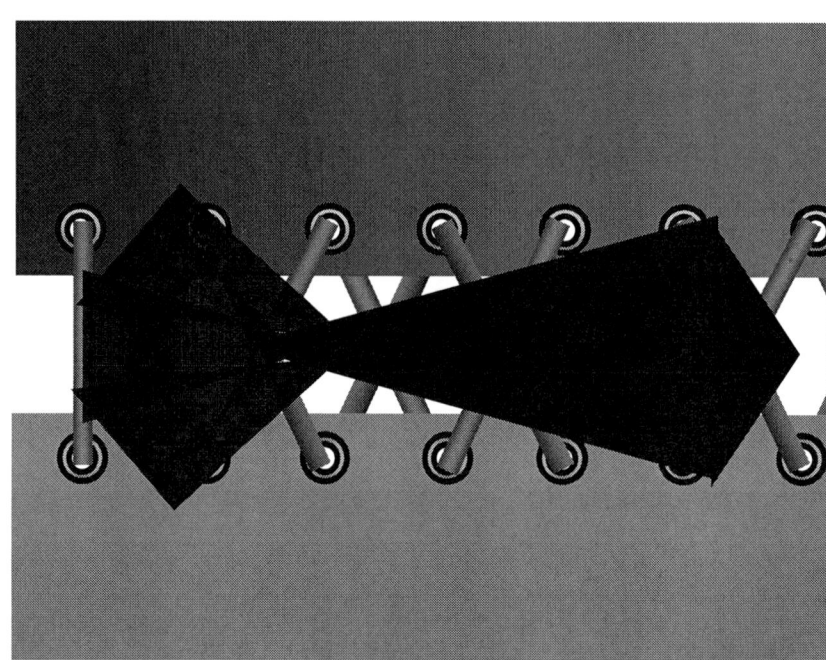

DFM is in Fashion. Wear it!

Based on recent Intel presentation*

* Trends in DFM, Sunit Rikhi, SEMI ISS, Jan 2007, Calif.

Chartered
semiconductor manufacturing

Design for Yield, Design for Manufacturing and Design for Test

Gerhard Fischer
Vice President, CHRT Europe
Chartered Semiconductor
4/30/07

© 2006 Chartered Semiconductor Manufacturing Ltd.
All rights reserved. No part or parts hereof may be reproduced, modified, adapted, distributed, republished, displayed, broadcast, transmitted in any manner or means or stored in an information retrieval system without the prior written permission of Chartered Semiconductor Manufacturing Ltd.

The Challenge - Today's Product Cycles
Steeper product ramps, shorter dwell in market

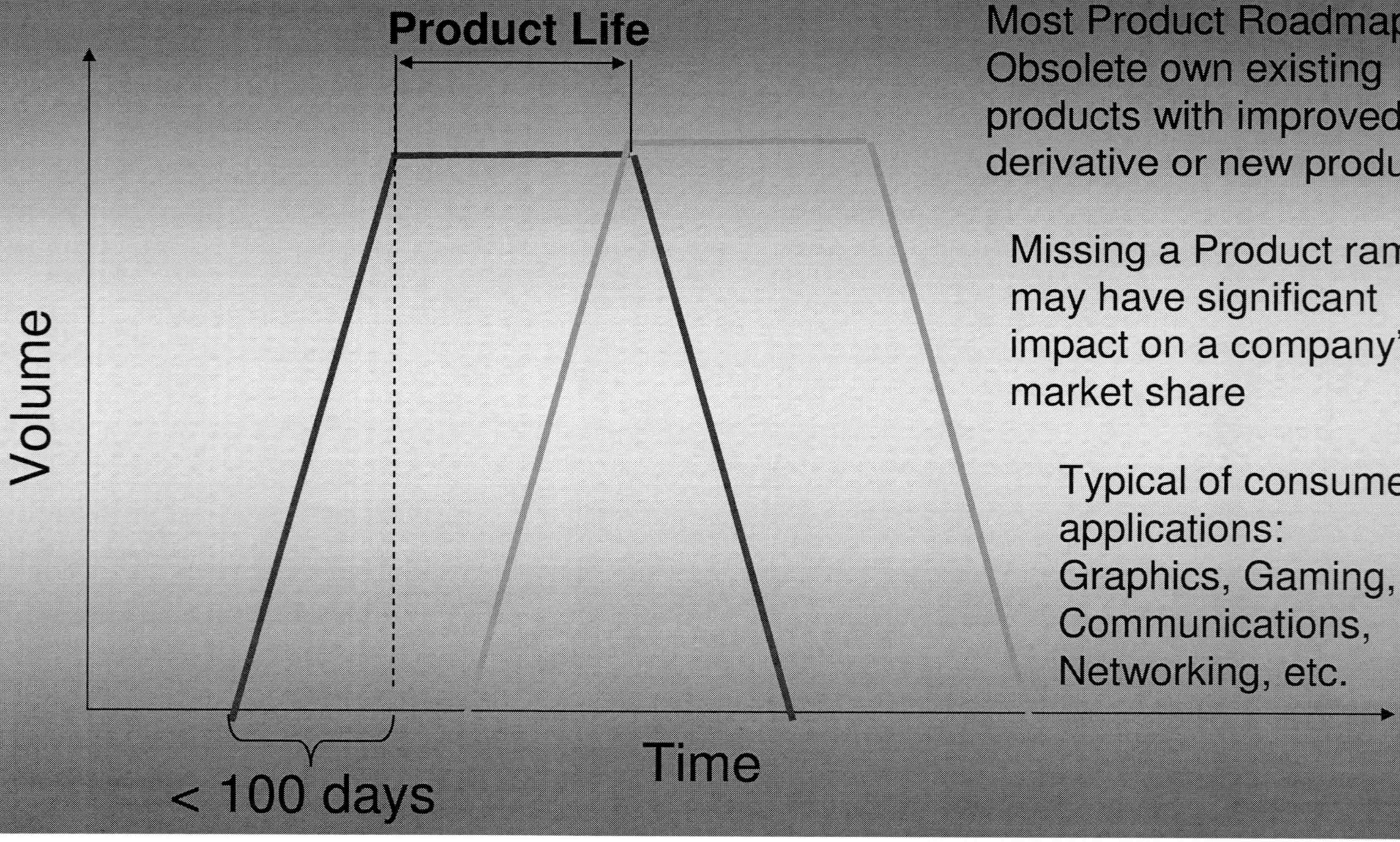

Most Product Roadmaps - Obsolete own existing products with improved derivative or new product

Missing a Product ramp may have significant impact on a company's market share

Typical of consumer applications: Graphics, Gaming, Communications, Networking, etc.

Product Life

Time

< 100 days

Volume

Design to Manufacturing to Test
COT Model Challenge

Design → **Manufacturing** → **Testing**

RTL Synthesis	Design Planning	Block Design	DFT ATPG	Chip Assembly	Physical Extraction	Verification	Mask OPC ORC	Proto-type Tape out	Testing	Yield Analysis	Failure Analysis

Design for Manufacturing / Design for Yield

DFT

DFT

Design to Manufacturing to Test Hand-off
- Coordination more critical than ever before
- More Manufacturing and Testing concerns addressed as part of Design

DFY / DFM
- Manufacturing Models being made available to customers
- Tools and Technologies being validated and customer enabled
- Manufacturing tools and control improving significantly

DFT
- Traditional Test Insertion done as part of design
- As a part of Test and Failure Analysis, DFT needs to be able to support Faster Yield Ramps

Chartered
semiconductor manufacturing

Common Platform DFM Initiative
Collaboration Among Industry Innovators

BLAZE

clear shape technologies

cadence

MAGMA

Mentor Graphics

ponte
the design-for-yield company

SYNOPSYS®

DFM Model Kit

Critical Area Analysis
CMP Simulation
Shape Simulation
Variation Aware Timing
Simultaneous Leakage Reduction & Yield Optimization

DFM Rule & Utility Kit

DFM Layout Guidelines
DFM Router Enhancements
DFM Checking Decks

Design Enablement Kit

Interconnect Modeling
DRC / LVS
DFM Services

Design for Manufacturing / Yield

RTL Synthesis
Design Planning
Block Implementation
DFT / ATPG
Chip Assembly
Extraction
Physical Verification
Mask/ OPC / ORC

Library

Power, Timing & Signal Integrity

Chartered
semiconductor manufacturing

Manufacturing Focus on Yield

Large scale automated Integrated Yield System (IYS) with real-time wafer-level tracking for fast yield learning and turnaround

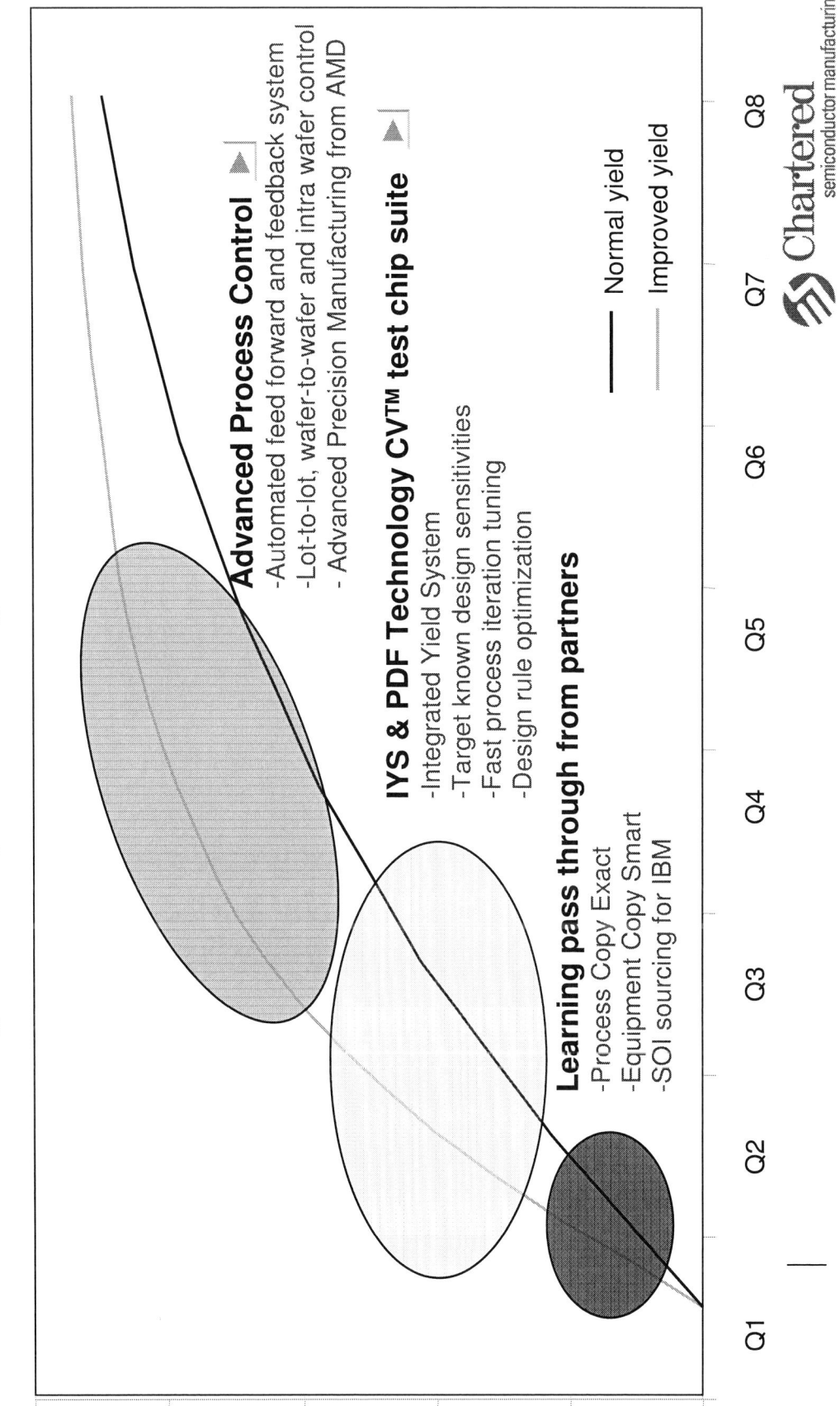

Advanced Process Control ▲
- Automated feed forward and feedback system
- Lot-to-lot, wafer-to-wafer and intra wafer control
- Advanced Precision Manufacturing from AMD

IYS & PDF Technology CV™ test chip suite ▲
- Integrated Yield System
- Target known design sensitivities
- Fast process iteration tuning
- Design rule optimization

Learning pass through from partners
- Process Copy Exact
- Equipment Copy Smart
- SOI sourcing for IBM

— Normal yield
— Improved yield

% Line Yield

Q1 Q2 Q3 Q4 Q5 Q6 Q7 Q8

Chartered
semiconductor manufacturing

Today's State-of-the-Art Foundries
Fab7 Automation Architecture

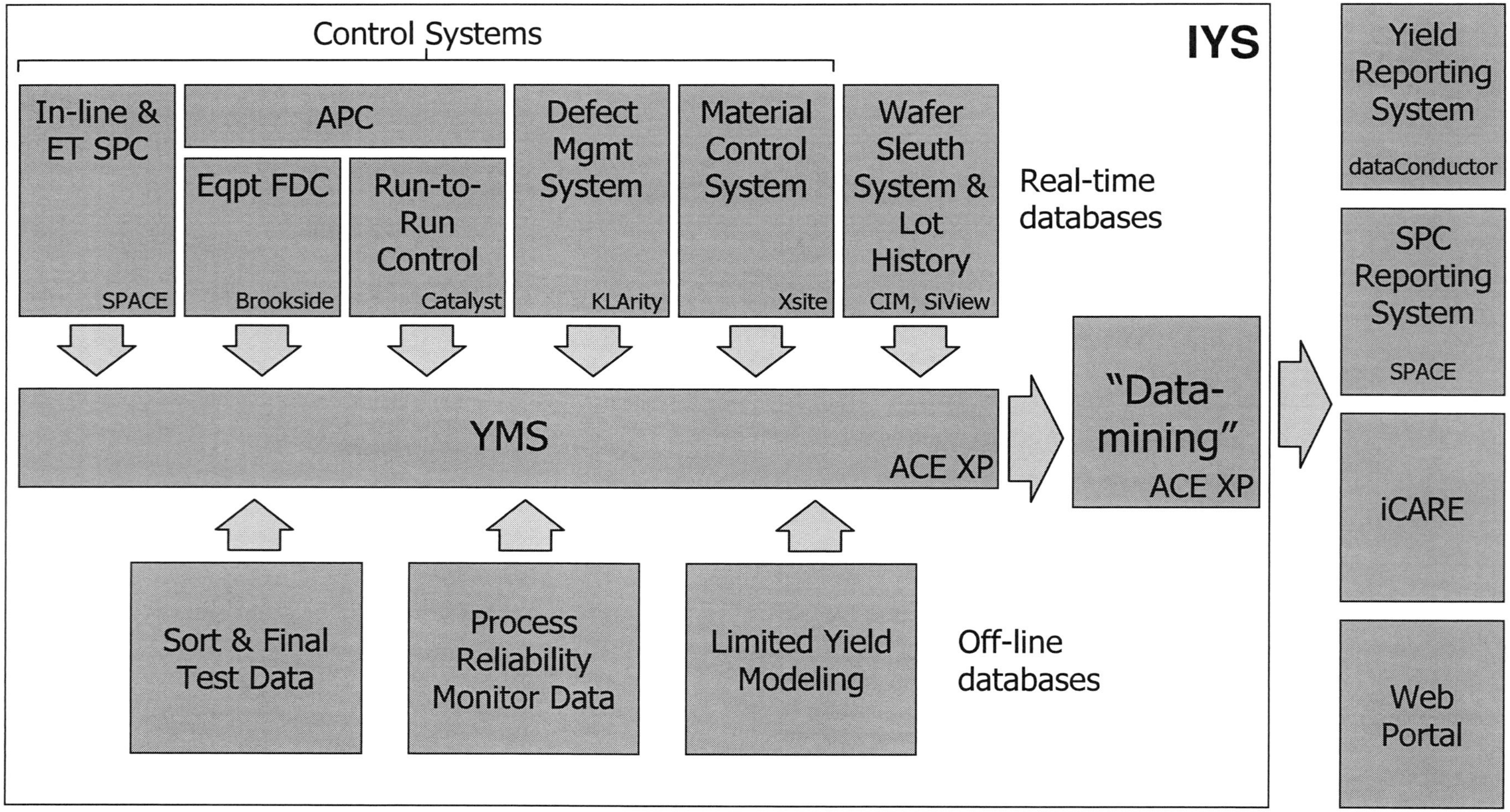

Yield Reporting System — dataConductor

SPC Reporting System — SPACE

iCARE

Web Portal

IYS

Control Systems

In-line & ET SPC — SPACE

APC / **Eqpt FDC** — Brookside / **Run-to-Run Control** — Catalyst

Defect Mgmt System — KLArity

Material Control System — Xsite

Wafer Sleuth System & Lot History — CIM, SiView

Real-time databases

YMS — ACE XP

"Data-mining" — ACE XP

Off-line databases

Sort & Final Test Data

Process Reliability Monitor Data

Limited Yield Modeling

Chartered
semiconductor manufacturing

Process Characterization
PDF CV™ Test Chip Suite

BEOL CV (CA,M1,Mx,Mx)
- CA
- Me
- Pri
- To
- Co
- BE
- Pull an
- pdFasT

FEOL 2 CV (AA to M1, M2, Mx)
- MC – CA Ma
- Salicide
- Printability e
- Other Syster
- Reduced Def (and M2)
- Pull and test
- pdFasTest
- Parametric (A

FEOL 1 CV (AA to M1)
- Poly, AA def
- CA defectivi
- MC defectivi
- Contact Sta
- *Very Limitec*
- Pull and test
- pdFasTest
- Parametric (A

Poly CV (AA and PC)
- PC defectivity (incl. triple gate oxide)
- STI topography
- STI seam
- PC ECD variations
- Salicide resistance
- Printability evaluation (PC)
- Pull and test after salicide
- pdFasTest

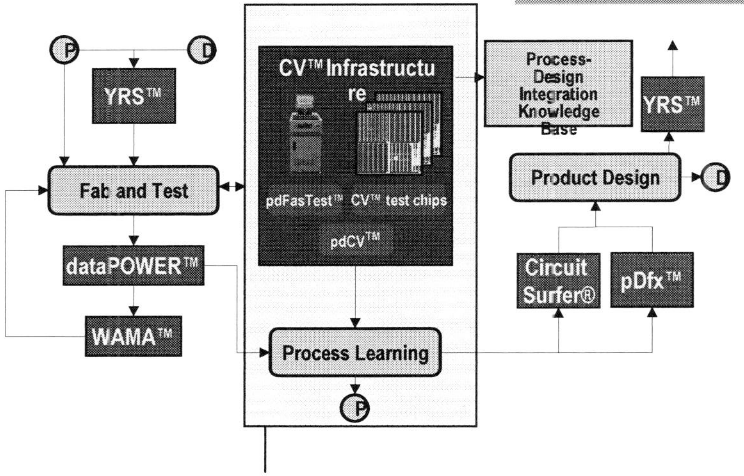

- Build yield infrastructure using PDF CV
- Characterization vehicles advantages:
  - Short cycle time in process feedback
  - Able to identify process issue
  - Enhance process margin

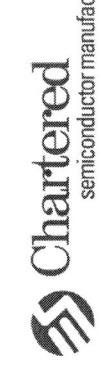

Chartered
semiconductor manufacturing

Advanced Process Control (APC)
Optimizing Process Efficiency and Dependability

■ Real-time process optimization

■ Tighter control of process parameter by automated real-time feed forward and feedback system

■ APC can apply to lot-to-lot, wafer-to-wafer and within wafer control

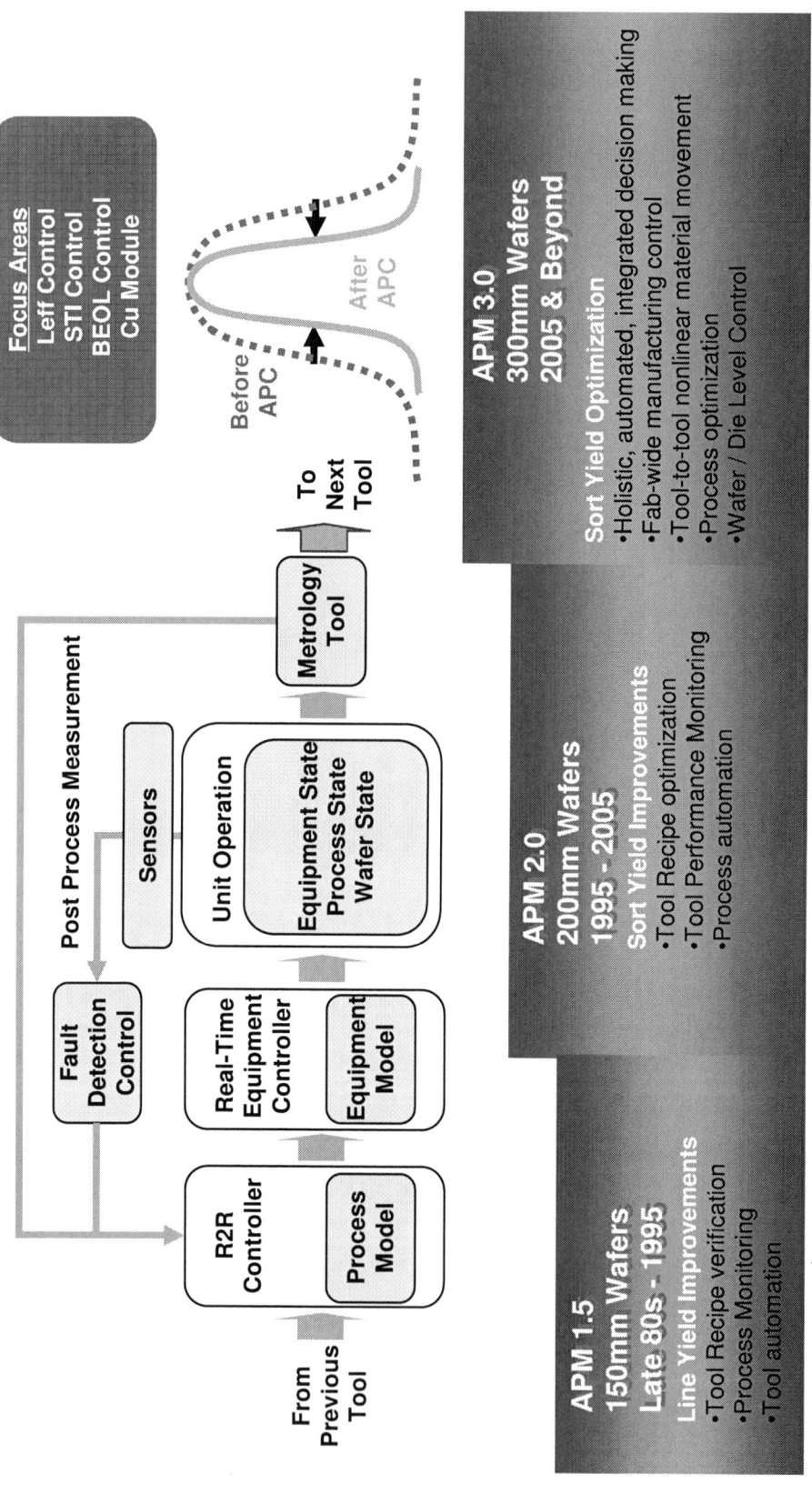

Focus Areas
Leff Control
STI Control
BEOL Control
Cu Module

Before APC
After APC

To Next Tool

Metrology Tool

Post Process Measurement

Sensors

Fault Detection Control

Unit Operation
Equipment State
Process State
Wafer State

Real-Time Equipment Controller
Equipment Model

R2R Controller
Process Model

From Previous Tool

APM 1.5
150mm Wafers
Late 80s - 1995
Line Yield Improvements
•Tool Recipe verification
•Process Monitoring
•Tool automation

APM 2.0
200mm Wafers
1995 - 2005
Sort Yield Improvements
•Tool Recipe optimization
•Tool Performance Monitoring
•Process automation

APM 3.0
300mm Wafers
2005 & Beyond
Sort Yield Optimization
•Holistic, automated, integrated decision making
•Fab-wide manufacturing control
•Tool-to-tool nonlinear material movement
•Process optimization
•Wafer / Die Level Control

Chartered
semiconductor manufacturing

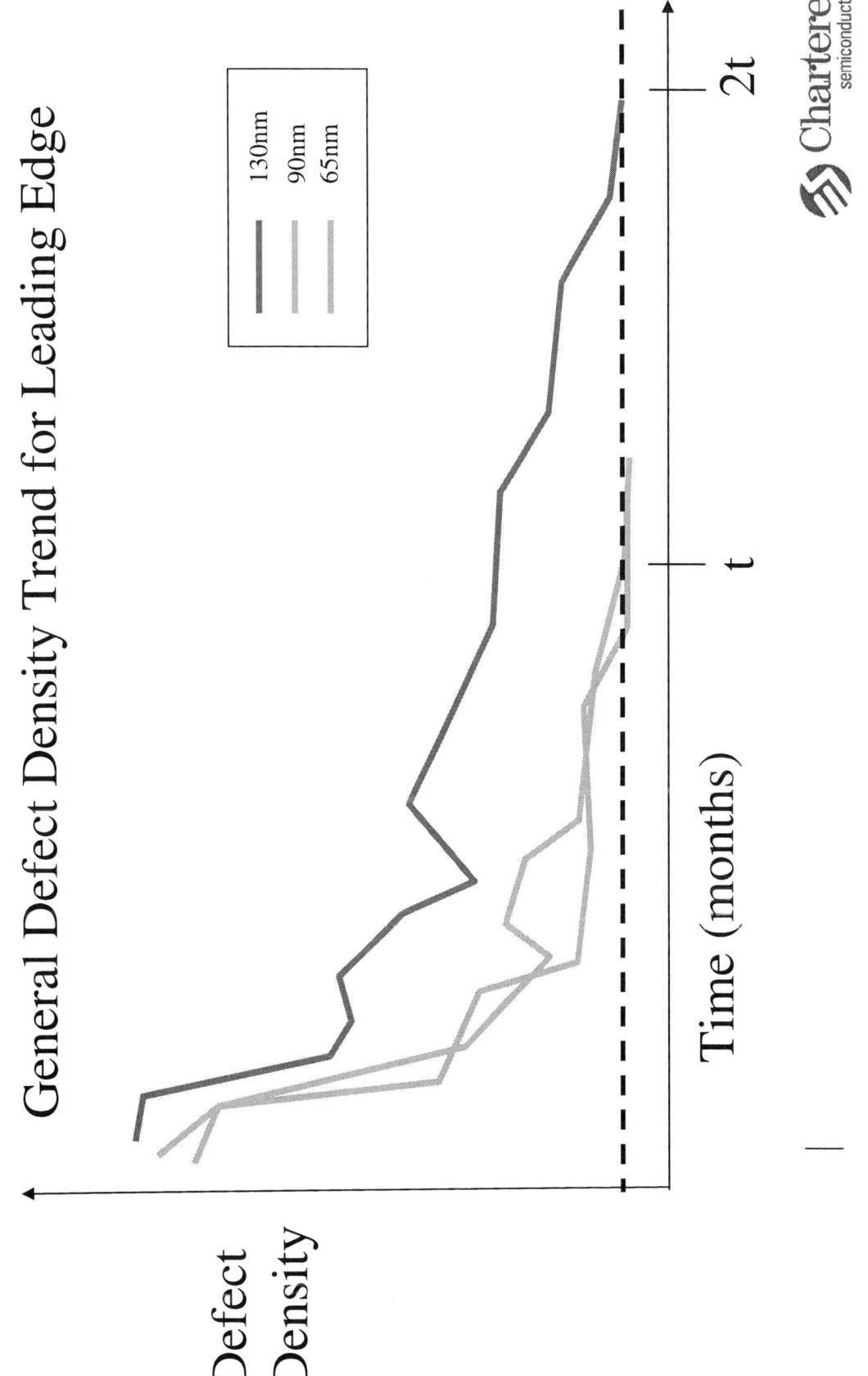

Intelligence into DFT / Yield Ramp

Fab Manufacturing

Yield Detractors Manufacturing Model

Data from CV's should be used to improve fault diagnostics and thereby reducing time to yield ramp

Yield Detractors Manufacturing Model

Data from CV's used to populate data models for various tools

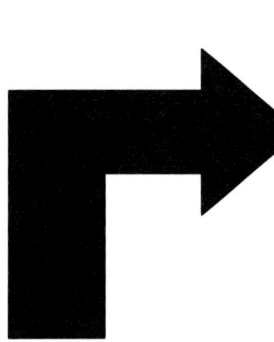

CV™ Infrastructure

pdFasTest™ CV™ test chips

pdCV™ Layout Dependant Models
- Fail Rates
- Performance Variability
- Module Sensitivities

Design for Test / Yield Ramp Enablement

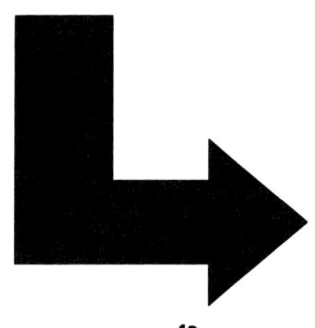

Test

Failure Analysis

ATPG

Fault Diagnostics

Manufacturing

Design for Manufacturing / Yield Enablement

Design

BIST & Test Insertion

 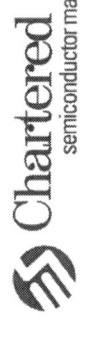 Chartered
semiconductor manufacturing

Summary

- One of the key challenges in supporting leading edge tape outs is requirement for very fast product yield ramp. This requires as much of a "first-time right" methodology and very fast test, failure analysis and repair capability.

- Progress is being made in the DFY/DFM area. The Common Platform has been one of the leaders in the industry in verifying and enabling promising tools and technologies in this area.

- Despite the challenges of leading edge manufacturing, Chartered has been able to continue world class manufacturing capability with the pass through learning from our joint development work with our partners, our IYS system, the PDF CV's and the Advanced Process Control system including AMD's Advanced Precision Manufacturing.

- Manufacturing at 90nm and 65nm has allowed Chartered to demonstrate world class defect censities in half the time it took to drive 130nm defect densities to the same level.

- DFT or rather the combined area of DFT/Yield Ramp has taken a new importance and one which could help to address the significant challenges if Test were to be made slightly more intelligent with information from manufacturing.

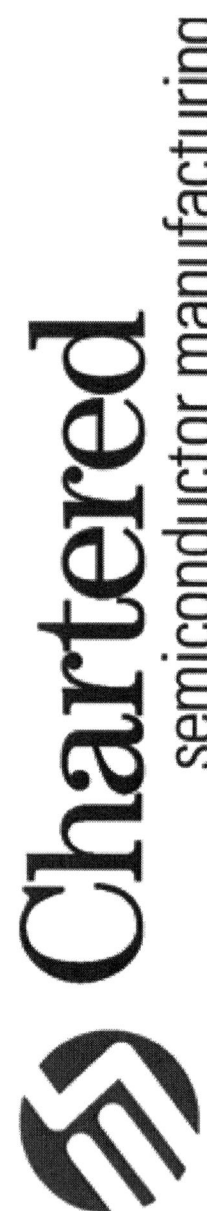

IET & FSA International Semiconductor Forum

Paris, May 15th 2007

DFT, DFY and DFM: Ways to build
Quality from the outset

Luc Tissot
European Product Specialist Mgr

Mentor Graphics®

The History of IC Yield Paradigms

DFM Recommended Rules

Random	Systematic	Parametric
Min Spacing	Max or Min Area	Poly Gate Extension
Min Width	Width/Space Comb	L-Poly to Active
Via Redundancy	Min Overlap of Via	Min Gate Width

Lithography-aware Design

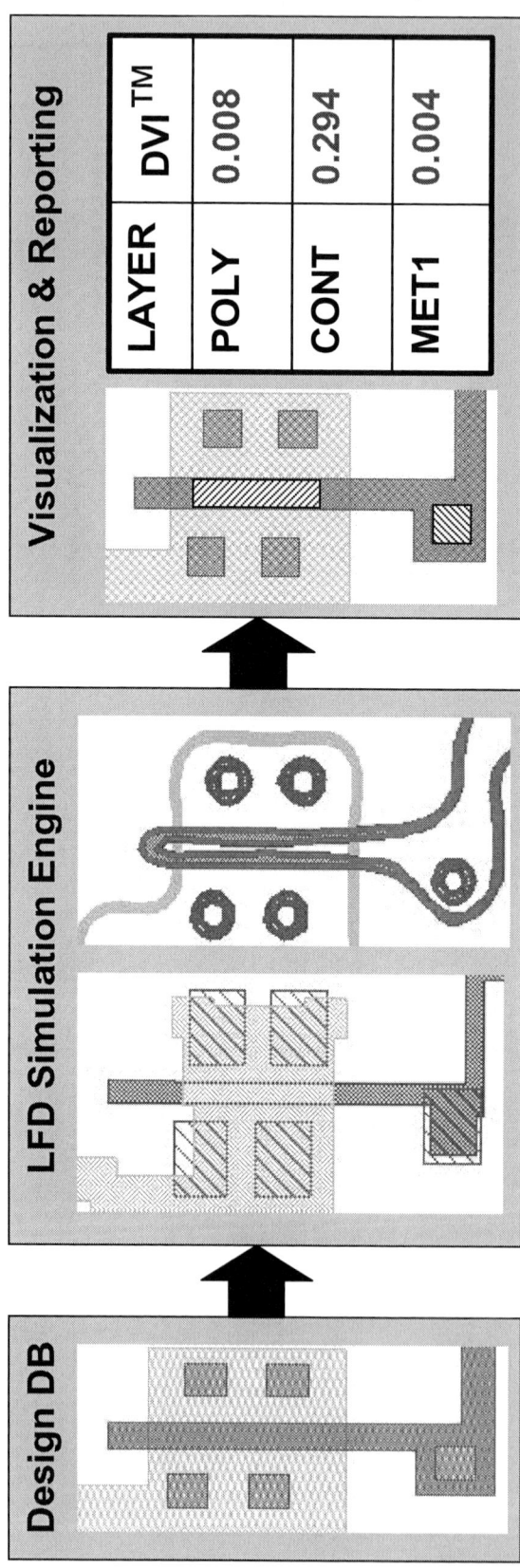

LAYER	DVI™
POLY	0.008
CONT	0.294
MET1	0.004

Automated Layout Enhancements

(A) Via Doubling

(B) Symmetric Via Doubling

(C) Via Enclosures & Line End Extensions

(D) Min Area Grow

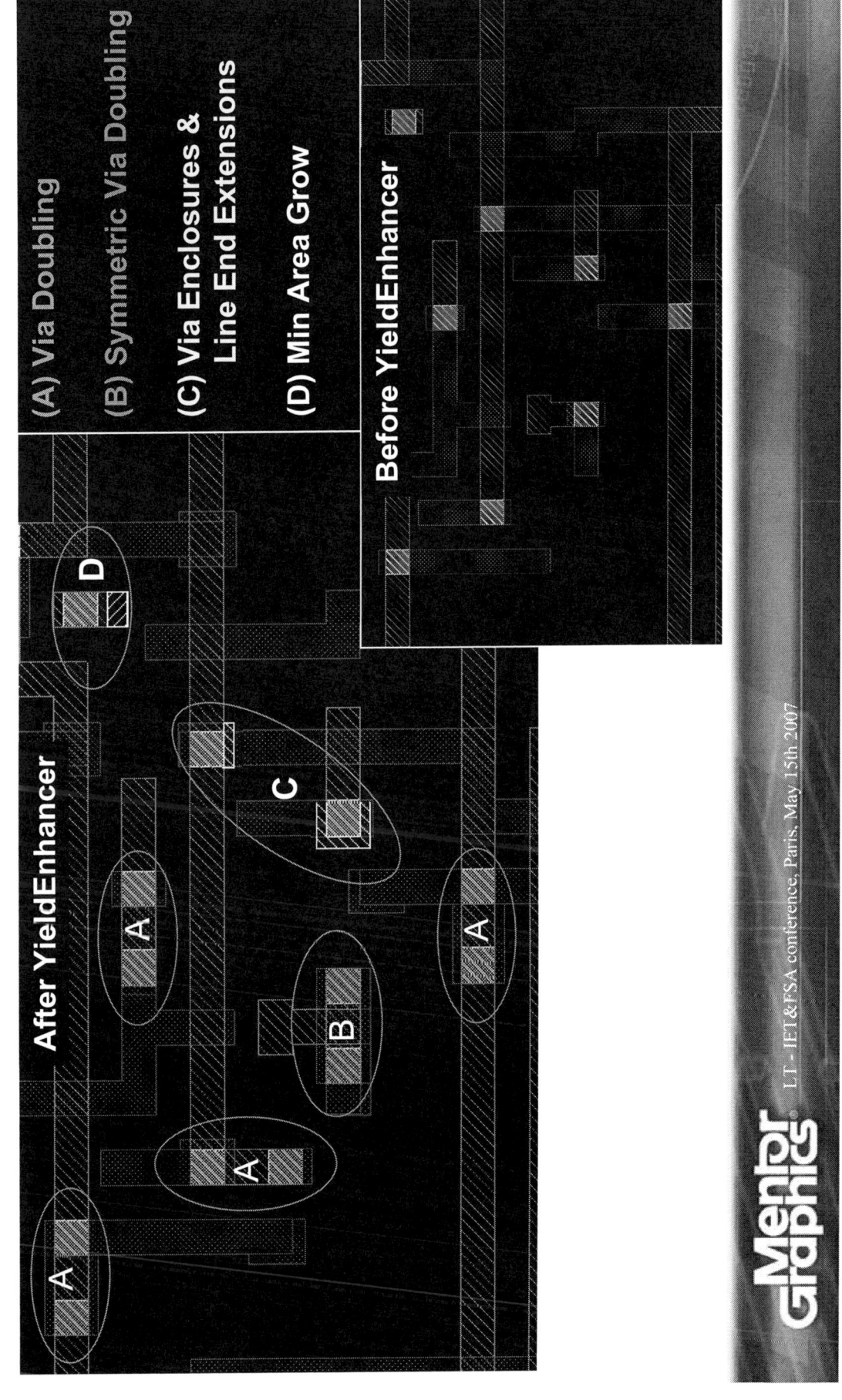

After YieldEnhancer

Before YieldEnhancer

Improved Test Quality: Physical Test

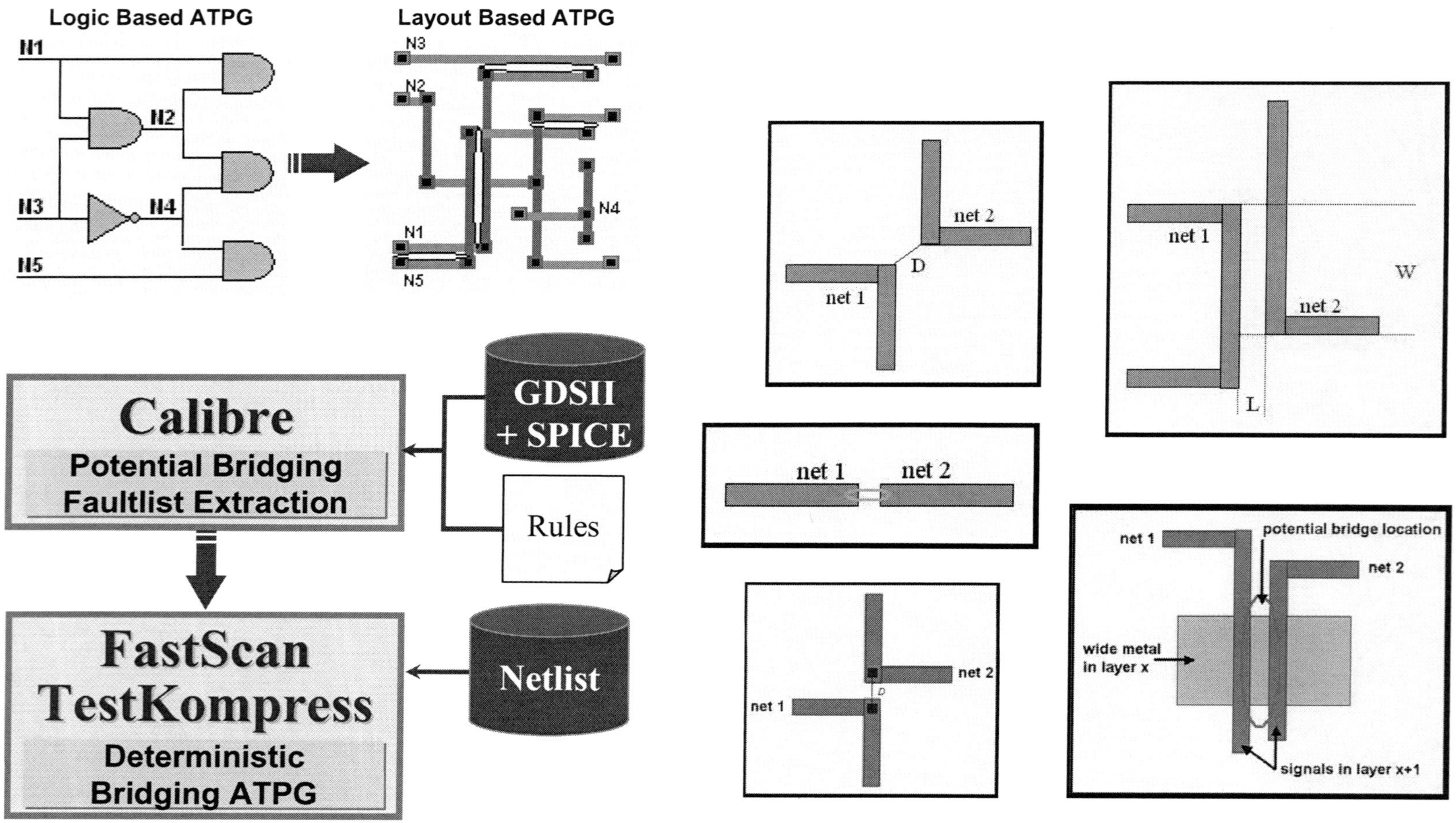

LT – IET&FSA conference, Paris, May 15th 2007

Improved Learning: Physical Diagnosis

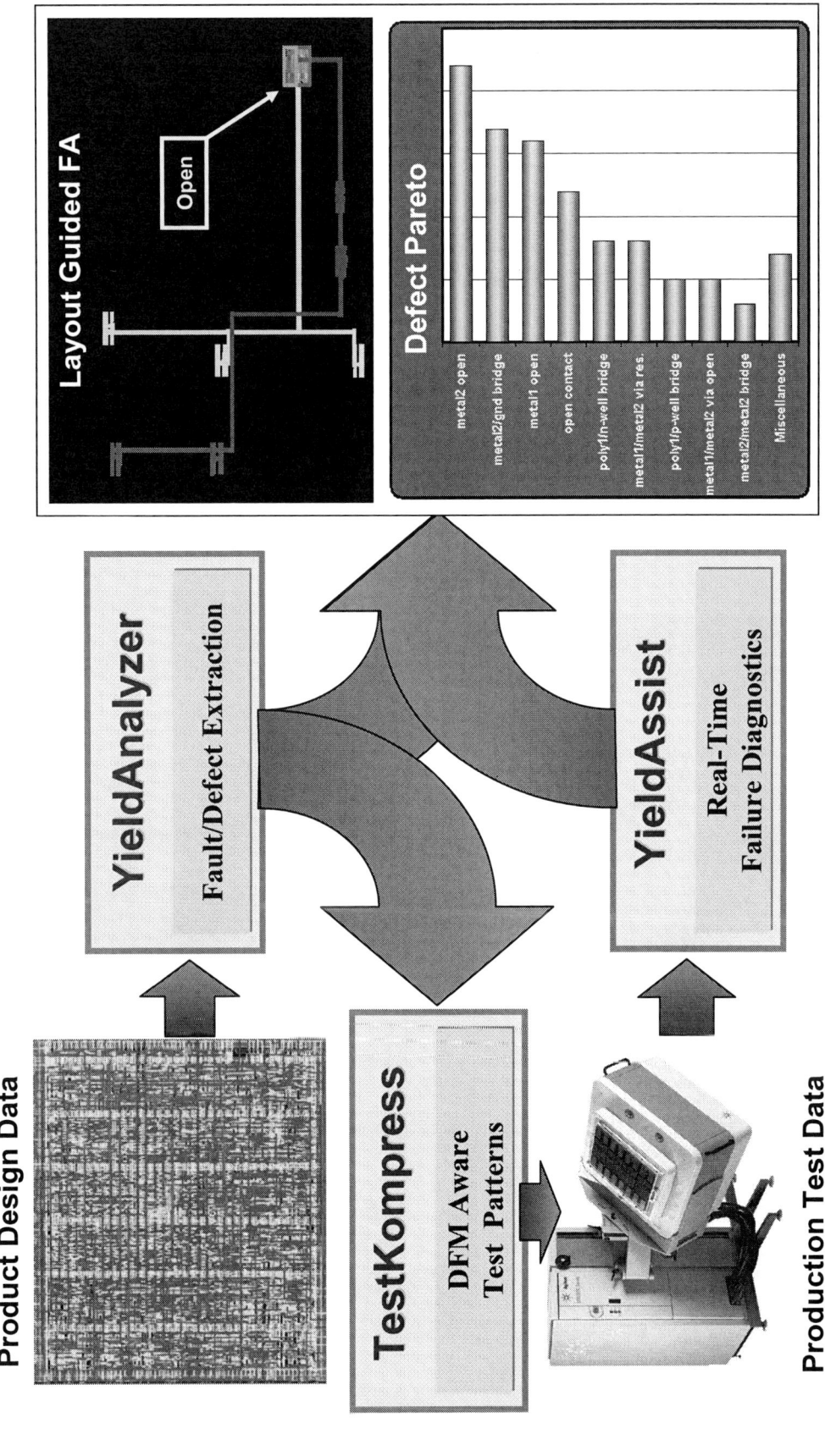

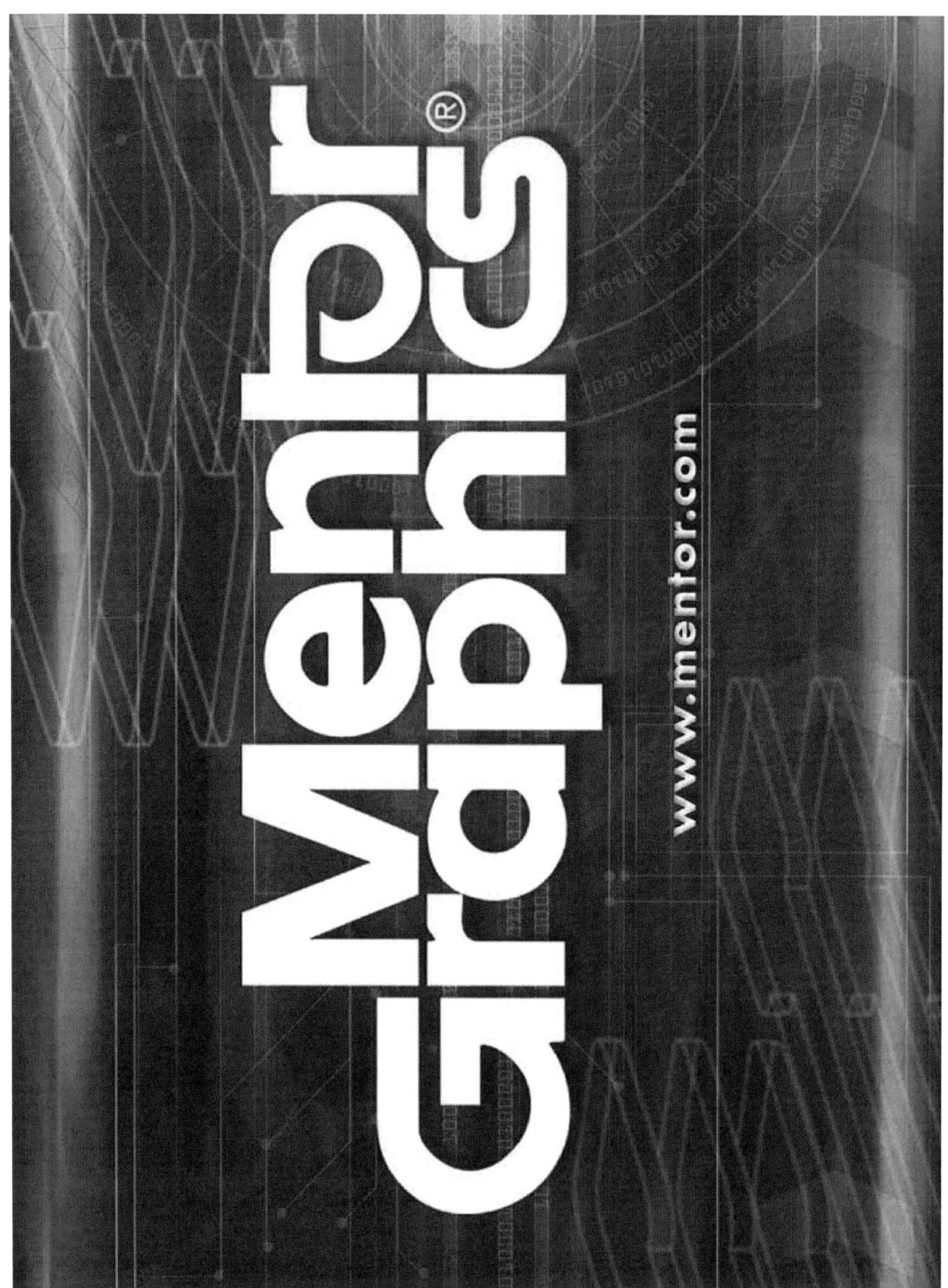

AUTHOR INDEX

Barratt, C. 39

Curren, G. 192

Dierickx, B. 44

Fischer, G. 205

Longford, A. 32

Nave, R. 101

Peeters, A. 148

Tissot, L. 217

Torten, R. 175

Wood, D. 65

Yu, J. 109